THE HISTORY OF THE EIGHTH BATTALION THE QUEEN'S OWN ROYAL WEST KENT REGIMENT
1914—1919

VICTORY!

By permission of] [*The Proprietors of "Punch."*

PUNCH'S VICTORY CARTOON USED FOR THE BATTALION CHRISTMAS CARD, 1918.

[*Frontispiece*

THE HISTORY OF THE EIGHTH BATTALION THE QUEEN'S OWN ROYAL WEST KENT REGIMENT

1914—1919

TO ALL THOSE COMRADES, BRAVE MEN AND TRUE, WHO FOUGHT AND DIED, THIS BOOK IS REVERENTLY DEDICATED.

In Flanders fields the poppies blow
Between the crosses, row on row,
  That mark our place ; and in the sky
  The larks, still bravely singing, fly
Scarce heard amid the guns below.

We are the Dead. Short days ago
We lived, felt dawn, saw sunset glow,
  Loved and were loved, and now we lie
      In Flanders fields.

Take up our quarrel with the foe:
To you from failing hands we throw
  The torch ; be yours to hold it high.
  If ye break faith with us who die
We shall not sleep, though poppies grow
      In Flanders fields.

By permission of the Proprietors of "Punch."

INTRODUCTION

On the pages of this book an attempt has been made to set down the plain unvarnished story of the life of a Service Battalion during the European War of 1914–1919.

The book originated with a desire, strongly expressed by all ranks of the Battalion in December 1918, that a history should be written. Accordingly a Battalion History Fund was formed, and the work was commenced.

Many difficulties have been encountered since then which have delayed publication considerably.

The narrative, based on the official War Diary of the Battalion, kindly lent by the Officer-in-Charge of Regimental Records, is strictly correct in all main facts. A mass of other detail obtained from note-books and recollections of many who served with the Battalion has had to be sifted and analysed, and if there are criticisms of minor points, we can only assure readers that we have done our best to get accurate information. We regret that from the details available it has not been possible to prepare an accurate list of the total casualties, and we are therefore not attempting to give figures, which would be incomplete. In many cases, however, we have been able to give the number of casualties for definite actions or over certain periods.

The photographs were taken after the Armistice, and consequently show places in their condition at the close of the war.

The maps are included by permission of the Geographical Section, General Staff, who gave us every assistance to reproduce them at a reasonable cost.

The portraits are reproduced by the kind permission of the respective photographers.

We would like to acknowledge here the assistance and the considerate treatment received from the printers, Messrs. Hazell, Watson & Viney, Ld.

It is hoped that it will now be possible to distribute the history to those who served in the Battalion and who can still be traced, and to the relatives of our comrades who gave their lives.

We trust that all into whose hands the book may come will read and cherish it, with all its defects, as a memento of the days when they or those dear to them fought for England in the trenches of France and Flanders.

H. J. W.
H. S. B.

March 1921.

CONTENTS

PHASE VI

PHASE VII

PHASE VIII

PHASE IX

PHASE X

ILLUSTRATIONS

ILLUSTRATIONS

ILLUSTRATIONS

MAPS

8th BATTALION, THE QUEEN'S OWN, ROYAL WEST KENT REGIMENT

PHASE I

THE FORMATION OF THE BATTALION—TRAINING IN ENGLAND—OVERSEAS TO FRANCE—THE BATTLE OF LOOS

September 1914—*September* 1915

THE formation of the 8th Battalion, The Queen's Own, Royal West Kent Regiment may be said to have commenced early in September 1914. The men who enlisted at the various recruiting offices of the Royal West Kent Regiment at that time were sent down to the regimental depot at Maidstone. The depot was not built to accommodate half the number of recruits which came pouring in, and the new-comers got a very bad impression of what the Army would be like. The feeding arrangements were bad, and they found that a man was lucky if he got someone else's dirty plate, and still luckier if he got some food to put on it. After 5 p.m. all were allowed out into the town, and it was there that the majority got their only real meal of the day.

The sleeping accommodation was quite inadequate. There were about a dozen tents, and the gymnasium, dining-hall, and all available spaces were crowded. But even then many men had to sleep out of doors. A few slept in the town, as, although no one was allowed out of barracks, the military police on duty let it be known that it was possible to get through the railings without being seen.

The average stay at Maidstone was about three days, and on Saturday, September 12, some 600 or 700 men were collected and taken by train to Shoreham-by-Sea, under the command of Major Venables, who returned at once to Maidstone. The reception at Shoreham was even less encouraging than at Maidstone. After considerable difficulty the men were got into some sort of order and marched in simply pouring rain to a field opposite the infirmary, where there was no accommodation at all, and they were told to stay under the trees to keep dry. By this time every man was wet through, and as there was apparently nobody present with much authority and no likelihood of an improvement, it is small wonder that many men took train for home, returning on Monday, when things were rather better. For those that remained bell tents were eventually pitched, the men being told off ten to a tent. Later some wet blankets and equally wet loaves of bread were distributed. No feeding utensils were issued for some days, and a piece of wood or newspaper had to serve as plate, and the fingers as knife, fork, and spoon. The tea invariably had half an inch of grease on the top, while the food was almost uneatable.

PHASE I

On Monday, September 14, the majority of those who had gone away on the Saturday returned, and a further detachment of some three hundred arrived, bringing the number up to nearly one thousand. Up to this time there seems to have been no officer present at all, and the Battalion had been under the command of Sergt.-Maj. Evans, but on the 14th Col. E. Vansittart, Commanding Officer, and Lieut. and Quartermaster Guess arrived, and things began to improve. Capt. Middleton arrived on the 17th, and took up the duties of Adjutant.

The difficulties of maintaining discipline will be understood when it is said that there were now in camp at Shoreham the infantry of a whole division, all in mufti, and all, with the exception of a very few officers and a few N.C.O.s, absolutely untrained. The only man in the Battalion in uniform was Sergt.-Maj. Evans. The other N.C.O.s were recognised by a red arm-band for sergeants and a white arm-band for corporals, while an officer was only distinguishable by the carrying of a walking-stick. The number of experienced N.C.O.s was quite inadequate, and the necessary establishment had to be made up by picking out the men who seemed most intelligent and likely to act as leaders.

The Battalion was brigaded with the 8th Battalion, The Queen's, Royal West Surrey Regiment, the 8th Battalion, The Buffs, East Kent Regiment, the 9th Battalion East Surrey Regiment, and for the first few weeks the 13th

Battalion Royal Fusiliers, which made up the 72nd Infantry Brigade, under the command of Brig.-Gen. B. R. Mitford, C.B., D.S.O. This Brigade, with the 71st and 73rd Brigades, formed the infantry of the 24th Division, under Maj.-Gen. Sir John Ramsay, K.C.B.

By the end of the month the camp was moved to the Oxenfield, or Recreation Ground, and the commands in the Battalion were then as follows :

Commanding Officer, Col. E. Vansittart.
2nd-in-Command, Maj. L. Brock-Hollinshead.
Adjutant, Capt. C. de C. Middleton.
Quartermaster, Lieut. C. Guess.
Sergt.-Maj. H. Evans.
Quartermaster-Sergt. W. Dadd.
Officer Commanding A Company, Maj. F. M. Walters.
Colour-Sergt. Skinner.
Officer Commanding B Company, Maj. E. V. O. Hewett.
Colour-Sergt. Lee.
Officer Commanding C Company, Maj. O. J. Daniel.
Colour-Sergt. Fisher.
Officer Commanding D Company, Maj. J. C. Chillingworth.
Colour-Sergt. Burn.
And about four subalterns per company,

Training was, of course, very quickly in full swing. It consisted for the first few weeks of squad drill and marching and nothing else.

Photo by] [*Elliott & Fry, London.*

COLONEL E. VANSITTART, D.S.O.

Commanding the Battalion September 1914—September 1915.

Squad drill from 6.30 a.m. until 8 a.m., squad drill and a little marching from 9 a.m. to 1 p.m., and squad drill from 2 to 5 p.m. It became rather monotonous, but was, of course, very necessary. After a while a little physical drill was added as a slight variation, and just occasionally an afternoon was employed in marching down to the sea for bathing. On Saturdays work stopped at 4 p.m. instead of 5 p.m., while even on Sundays the 6.30 to 8 a.m. parade was held in addition to church parade, so that it can be seen that we were starting in no half-hearted fashion. So seriously indeed were we setting out on our training, that officers did not dare to ask for permission to go to Brighton after hours, unless it was to get their hair cut, and it must be admitted that hair seemed to grow rather quickly in those days.

By this time the interior economy of the Battalion was more or less on an organised basis, and considerable progress began to be seen in all directions. In physique and keenness a very large majority of the men were splendid, and there was no doubt that we had the nucleus of a very fine battalion. On the other hand, there were some men who were obviously medically unfitted for an army life, and who ought not to have been passed in at all. Fortunately it was quite easy in these early days to discharge such men, and by the end of October those "unlikely to make efficient soldiers" had been pretty well weeded out. No one could have wished for a finer set of men than the 1,026 who were retained.

It was about this time that the blue uniforms were issued. They were blue serge, and though not beautiful and not particularly military, they gave the Battalion an appearance of uniformity, and were an enormous improvement on the very varied collection of mufti previously in use. In addition to appearance, the change was very beneficial to discipline, as it had been formerly impossible to tell whether a man was a soldier or not.

The month of October was brilliantly fine, and very great progress was made with the training, but November was not so kind, and the weather got worse and worse, until by the end of the month the camp was no better than a mud patch about six inches deep, and, as always happens in a muddy camp, had become very unhealthy, and was condemned by the sanitary authorities.

The Division then went into billets at Brighton, Shoreham, and Worthing. The 72nd Brigade went by train on November 30 to Worthing, where the Battalion was billeted at the western end of the town, and had very good billets indeed. The people of Worthing were most kind to us, and no one will forget the very pleasant four months' stay there. Right up to the end of the war we continued to get regular parcels of comforts sent out by the townspeople, and we cannot be too grateful to the town of Worthing for their kindness both during our stay and afterwards.

The training facilities were perhaps not quite so good at Worthing, but the Downs could still be reached by a short march, and the Recreation Ground was excellent for close-order drill parades.

One of the difficulties at Shoreham had been the absence of rifles. At first only a few imitation rifles made of solid wood were available; then a little later six D.P.s (i.e. for drill purposes only, unsafe for firing) were issued per company, and were used for training the officers and N.C.O.s in musketry. Shortly after reaching Worthing, D.P.s were issued to the whole Battalion, and arms drill became for a time one of the principal items on the programme of work.

The extreme, and in fact excessive ardour with which we had entered into our training had been wisely tempered down, and parades now lasted about seven hours a day, instead of eight and a half; we also had a proper half-holiday on Saturdays, and the early morning parade on Sundays had been knocked off. Christmas leave of four clear days was given to the whole Battalion in batches between the third week in December and the second week in January. When this was over, company training was started, and this was the order of the day for the rest of the stay at Worthing.

We now had supplies of full-dress scarlet tunics, which were obtained from the depot, and which were worn for church parades and when proceeding on leave. There were not enough to go round the whole Battalion, only 460 coats and trousers and 420 caps, so the men who appeared on parade in the ordinary blue were either excused or marched to church as a separate party.

About the beginning of March the Battalion was issued with its first khaki service dress, and this was a most cherished possession, and was only worn for battalion drill parades, which took place each Saturday, and for church parades on Sunday, the scarlet being then returned to Maidstone. We had previously, however, about the beginning of January, been issued with khaki greatcoats, which had been taken into general use. Five miniature rifle ranges were erected at Worthing, and much morris-tube firing was done, and the musketry made great progress all round.

At the end of March the Battalion moved by train from Worthing, which was left with many regrets, to Redhill, where we were again in good billets. The only work or training done during the three weeks' stay there was digging trenches for the London defences at Merstham. This meant parading at about 7.30 a.m. and marching four and a half miles, working until 3 p.m., with half an hour's break for eating haversack rations, and then marching home, a somewhat exhausting job and very tedious when repeated day after day.

In the third week of April the Battalion returned by train to Shoreham-by-Sea, and went into a new hutment camp which had been erected on the southern slopes of the Downs, just north of the village. The accommodation was quite up to the normal, except that we had by this time amassed something like seventy officers (every platoon had at least three, and most platoons four), and as a consequence the officers' accommodation, which was only meant for a normal establishment, was rather overcrowded.

From the point of view of training it was much more satisfactory to be back in camp again. Most of this stay at Shoreham was devoted to battalion training, and no more suitable piece of country could have been found. The only difficulty was that the whole division was now at Shoreham, and that in order to save congestion the ground was divided into areas and allotted weekly to units, which meant very often a long march before the actual training could be started. Usually the Battalion had covered some twenty miles as well as carrying out a tactical exercise by the time that it returned to camp in the afternoon. Night work by companies was carried out frequently.

Some of the transport wagons and animals were issued here, and cookers occasionally accompanied the Battalion. Many of the mules were untrained, and were taken in hand by Lieut. S. R. Paul and Sergt. Wood, up to that time Sergt. Master-Cook, who became transport officer and transport sergeant respectively. The majority of the transport, however, was not issued until we reached Aldershot.

The weather was, on the whole, very fine and hot, and although these long days in such weather were very tiring, there is no doubt that they got us into the very pink of condition.

The second suit of khaki was issued on arrival at Shoreham, and from that time was taken into general use, and we felt that we really were soldiers at last.

It was at Shoreham also that our band was formed from men who already had a knowledge of, and were able to provide, their own instruments. These were later supplemented by gifts from various officers. The band was taken in hand by Sergt. Baker, and was the pet hobby of Capt. Middleton, the adjutant. We had possessed our own drums for a long time.

At the end of June we moved by road to Blackdown Camp, near Aldershot. The march occupied three days, halts being made at Horsham and Guildford, where billets for the night were provided. Blackdown, though of course a satisfactory place from a purely military point of view, was by no means so pleasant in other respects as our former stations. It had a dirty appearance and the country was anything but beautiful. It also had the disadvantage of being a long way from a station or town, being almost equidistant from Aldershot, Farnborough, and Camberley.

The men were in huts and the officers either in the old married quarters of the barracks or else under canvas.

Soon after arrival at Blackdown service rifles were issued, and the bulk of the training was devoted to brushing up musketry prior to firing the general musketry course. Most of July and the early part of August were devoted to firing, and little else was done all through this period, as companies not actually firing were nearly always on duty in the butts. The results were rather disappointing, due very largely to the abominable weather. After the lovely days of the early part of the summer it was very unfortunate that we should have such bad luck for the musketry course. It became very much colder, the light was often most indifferent, and we were also bothered by frequent high winds and a lot of rain. On many of the days on which we fired, shooting would have been cancelled under peace conditions; but time was short, and we were forced to get on with the work if the targets could be seen at all. However, the course gave the men confidence in their new rifles, and a very large majority attained at least the standard of second-class shot in spite of the weather. Bombing instruction, under 2nd Lieut. Hough, and scouting, under 2nd Lieuts. Harris and Eccles, was also carried out, and the Lewis guns were issued and used for training.

On August 12 the Division was inspected by Field-Marshal Earl Kitchener, the Secretary of State for War. The whole Division was drawn up in one line of battalions in mass, and Earl Kitchener rode down the line from the right, and was received with the general salute. The whole proceedings, as far as each battalion was concerned, lasted about two minutes, but no one with experience of these functions will be surprised to hear that we were on parade waiting for over one and half hours before anything happened.

The rest of the month of August was devoted to training in trench warfare, final preparations for proceeding overseas, and embarkation leave. The trench-warfare scheme was carried out by the whole division. The two opposing forces marched under war conditions until they met on Chobham Common after dark, when each side commenced to dig in with entrenching tools until the large tools were brought up. Work was continued under war conditions until 7 o'clock the next morning, when peace conditions were resumed in order to enable the work of completing the trenches to be pushed on more quickly. It was intended that the troops should occupy the trenches for some days and that reliefs should be practised, and finally a trench-to-trench attack and consolidation carried out. It happened, however, that the weather was very bad, and the scheme was abandoned on the second day.

On that day the King and Queen visited the Division and walked along the trenches, speaking to many of the officers and men, and we felt that this foreshadowed a fairly imminent departure for other trenches. Immediately the trench-warfare scheme was abandoned, embarkation leave commenced, each man getting four clear days.

On about Wednesday, August 25, instructions were received to be ready to proceed overseas on the following Sunday. All leave was cancelled from Friday, and the last three days were spent in one long rush of packing up, kit inspection, exchange and issue of new kit, and all the necessary preliminaries for proceeding on active service.

On Sunday, August 29, the Battalion actually left Blackdown for Southampton. The organisation of the Battalion as it proceeded overseas was as follows:

Commanding Officer, Col. E. Vansittart.
2nd-in-Command, Maj. L. Brock-Hollinshead.
Adjutant, Capt. C. de C. Middleton.
Machine-Gun Officer, Lieut. W. K. Tillie.
Transport Officer, Lieut. S. R. Paul.
Quartermaster, Lieut. H. Evans.
Regt.-Sergt.-Maj. A. Lee.
Regt.-Quartermaster-Sergt. W. Dadd.

A Company:
Commander, Maj. A. H. Pullman.
2nd-in-Command, Lieut. R. M. Old.
Coy.-Sergt.-Maj. Campion.
No. 1 Platoon, Lieut. L. Gibbs.
No. 2 Platoon, Lieut. R. M. Old.
No. 3 Platoon, Lieut. H. S. Brown.
No. 4 Platoon, 2nd Lieut. C. D. N. Lawson.

B Company:
Commander, Capt. R. W. Grant.
2nd-in-Command, Lieut. P. T. Smith.
Coy.-Sergt.-Maj. Ackhurst.
No. 5 Platoon, 2nd Lieut. E. A. Bigsby (Billeting Officer).
No. 6 Platoon, Lieut. N. B. Green.
No. 7 Platoon, 2nd Lieut. V. G. Don.

No. 8 Platoon, Lieut. O. Jones (Engineering Officer).

C Company:

Commander, Capt. A. C. Edwards.
2nd-in-Command, Capt. C. A. Hutchinson.
Coy.-Sergt.-Maj. Mills.
No. 9 Platoon, 2nd Lieut. D. H. Watts.
No. 10 Platoon, 2nd Lieut. M. S. Ell.
No. 11 Platoon, Lieut. H. L. Lewis.
No. 12 Platoon, 2nd Lieut. A. N. Harris (Scout Officer).

D Company:

Commander, Major J. C. Chillingworth.
2nd-in-Command, Capt. P. M. Robertson-Ross.
Coy.-Sergt.-Maj. Acland.
No. 13 Platoon, 2nd Lieut. R. F. T. Burrell.
No. 14 Platoon, 2nd Lieut. H. O. Beer.
No. 15 Platoon, 2nd Lieut. G. de L. Hough (Bombing Officer).
No. 16 Platoon, 2nd Lieut. D. W. Plant (Signalling Officer).

We journeyed to Southampton in two trains, half of the transport going with each half battalion, and on arrival in the afternoon the Battalion was marched to the docks and there detained in a large shed, on account of submarines. The next day was spent in the shed, and the Battalion embarked late in the afternoon, and left the docks in time to reach the open sea, where the convoy was met by a patrol of destroyers, as it was getting dark. The sea, without being particularly rough, was sufficiently uncomfortable to make many people wish for dry land again. We anchored outside Havre in the early hours of the morning, and went into the harbour soon after daylight. There we disembarked, and marched up to camp on the high ground about Honkville. The troops were delighted with the new experience of being in a foreign country, and were in most excellent spirits. We remained in this camp until the next morning, when we marched down to the Gare des Marchandises and entrained at about 11 o'clock. This being the Battalion's first experience of travelling in cattle trucks (40 hommes and 8 chevaux), considerable disgust was expressed. But this mode of travelling became quite a normal thing for those who had an extended career in France. After a very slow journey of seventeen hours, via Rouen and Abbeville, we detrained at Montreuil at about 4 a.m. and marched to Hucqueliers, a halt being made on the way for breakfast.

The billets at Hucqueliers were very good, and the population was most kind to us. Training was continued by companies, considerable time being devoted to field firing, which was carried on in any likely spot, and seven ranges were constructed for target firing. Later some big divisional field days were included in the programme, the most memorable being an alarm in the night, followed by a field exercise lasting until 6 p.m. the next evening.

After a stay of nearly three weeks the Battalion

marched out of Hucqueliers at 6 p.m. on September 21, and proceeded to Dennebroeucq, which was reached in the middle of the night. There were no billets or camp available, and the men bivouacked in a field with no covering except their own waterproof sheets. There was a barn available for the officers, but it was so plentifully supplied with fleas that many wished they had remained in the field.

The next night the march was continued to Isbergues, the Battalion being billeted in the brewery, and a much more comfortable day was spent than the last. The next night a short move of about three miles was made and quite satisfactory barns and billets occupied; while on the night of the 24th the march was continued through Choques, which was full of guns, ambulances, and other very warlike appliances, to Bethune, where the Battalion was billeted in the tobacco factory, with the officers in an empty house on the opposite side of the road.

The four nights' marching had been rather exhausting, not so much on account of their length as on account of the congestion of traffic on the roads, which, combined with the darkness, made progress very slow.

The officers had already been told that they were going into the biggest battle the world had ever known, and were also told that their rôle would be to march through to Douai and hold it, a task which was considerably modified before we actually reached the battle.

On the morning of September 25 the 72nd

Brigade moved to Beuvry, and from there marched to a point behind Vermelles, where the battalions rested on the road-side. The Battle of Loos was now in full swing, and many prisoners and wounded were coming down from the line, but no very definite information could be got from them. The task now allotted to the Division was that it should pass through the village of Hulluch, which was believed to be in our hands, and then capture and organise the Pont-a-Vendin bridgehead some three and a half miles further east. As a matter of fact, the 73rd Brigade and two battalions of the 71st were detached to other parts of the line, leaving only the 72nd Brigade to carry out the attack.

When night fell the Brigade moved up from Vermelles, and opened out into artillery formation, as the exits to the village were being shelled, advancing first on Le Rutoire Farm and then on a compass bearing to the Lone Tree, where the Battalion halted and got into touch with the other battalions of the Brigade. Later the advance was resumed, with the Battalion leading the Brigade, and the final position reached during the night was a line of old German trenches parallel to and some 1,000 yards west of the Lens–La Bassee road. It is interesting to note that although this part of the line was supposed to be occupied, no troops were passed on the way up, and there were none in the trenches to be relieved.

Although the shelling had been fairly heavy, the casualties were not very serious, but included

Lieut. P. T. Smith, slightly wounded in the foot. He refused, however, to be sent down to hospital. Col. Vansittart was also slightly wounded in the hand during the night.

Our instructions were that we should attack at dawn, the objective for the Brigade being the main German position some 2,000 yards away, which stretched from Hulluch village on the left to the Bois Hugo and on to Hill 70 on the right, our frontage being from Hulluch to the Bois Hugo.

Considerable sniping and random shell-fire was kept up by the enemy during the night, but did very little damage. The Battalion scouts, under 2nd Lieut. Harris, were sent forward during the night, 25/26th, and came back with the report that Hulluch was in the hands of the enemy and strongly organised, and that they were also holding the line of the Lens–La Bassee road. The whole Battalion could not be accommodated in the main firing trench, and an attempt was made by C and D Companies to dig a parallel trench in front, but, owing to the approach of daylight, and the increasing attentions of the enemy's whizz-bangs and machine-gun fire, this was abandoned, and the surplus was withdrawn into a communication trench, where the men hollowed out scoops for themselves as additional protection from the shelling which was enfilading them.

No orders for the attack at dawn came to us, and we had to wait. As the light grew stronger, the enemy's artillery fire increased and casualties became more numerous, while our guns did heavy counter-battery work. At about 10.20 a.m. Col. Vansittart was sent for by the Brigadier and got orders to attack at 11 a.m. ; he reached the Battalion again at 10.53, and only had time to give orders for the whole Battalion to move up into the fire trench, and to tell off companies as firing line and supports. We were told that there were plenty more fellows in reserve to go through us, and that the troops on either flank would attack simultaneously with ourselves. We were to leave Hulluch village on our left, and it would be dealt with by the troops on our flank.

Punctually at 11 o'clock the first wave, with B and D Companies in the front line, left our trenches, followed at short intervals by the subsequent waves. The 9th East Surreys were on our right, and the 8th Buffs (right) and the 8th Queen's (left) were in support. Instead of a whole division attacking on our left, a handful of some fifty to seventy-five men, belonging to a Welsh regiment, alone left our trenches, the orders apparently having arrived too late, while the right flank of the East Surreys also appears to have been in the air.

The ground to be covered sloped gradually down to the Lens–La Bassee road, and from there rose to the main German positions, about 1,100 yards beyond, near the crest of the ridge running north from Hill 70. The attack was carried out at a marching pace in order to save the men's breath for the final charge and bayonet work, and the advance was so steady and the formations so regular that it looked more like a field day at Aldershot than part of a great battle. The enemy brought a very heavy rifle, machine-gun, and artillery fire to bear on the attacking lines, and shortened his range to keep pace with the advance with wonderful precision ; but the casualties up to this point, though severe, were not inordinately heavy.

When the first wave was still some hundred yards short of the main road, the enemy withdrew, some retiring in very open order up the hill, and many of these were shot down by our men, while stronger parties made off to either flank, those on our left gaining the village of Hulluch, and those on the right reaching a sunken road on the near side of the Bois Hugo. A very severe machine-gun fire from either flank swept the road, and very heavy casualties were sustained in crossing it. About 400 yards east of the road the Battalion passed a forward German trench full of enemy dead.

As the advance swept up the long slope towards the main German trenches the machine-gun fire increased in intensity, being very severe from both flanks. Shelling directed from behind Hill 70 and from the south-eastern part of Hulluch was also very heavy, and two field guns were brought up to a position from which they took the attack first from the flank and later from the left rear, firing practically over open sights. In spite of the very heavy casualties which were being sustained, the Battalion continued its advance up the slope with the utmost gallantry, but it was found on reaching the enemy's main position that the barbed wire protecting it was absolutely intact. What more hopeless situation could possibly have arisen ? When the orders for the attack were issued, it had been assumed that all the enemy's wire had been cut by our bombardment, but here, after a most gallant advance of about 2,100 yards in face of a murderous fire, not only from the front, but also from both flanks, the Battalion, and in fact the whole Brigade were faced with a thick belt of uncut wire, swept by artillery, rifle, and machine-gun fire from three different directions. With assistance on our flanks to keep the enemy occupied and to draw a due proportion of his fire, it would perhaps still have been possible to force a way through the wire and capture the trench. We wish to make it quite clear that we are not blaming our comrades on either flank for their failure to help us. The facts appear to be that at the time fixed for our attack the enemy himself launched a counter-attack against the troops on our right, who were consequently pinned to their ground, while the orders did not reach those on our left in time, with the consequence that the 72nd Brigade had both its flanks exposed, and the enemy was able to concentrate on the one brigade the fire of the artillery and small arms which would normally have been used on the front of three divisions. Many men made gallant attempts to get through the wire, 2nd Lieut. Don being killed when half-way through. The remainder dropped to the ground where they were and returned the heavy fire from the enemy's trench. The situation, however, was hopeless. In front was an impenetrable belt of wire, and the line, such as was left of it, raked with rifle and machine-gun fire from the front and both flanks, and enfiladed by an equally deadly artillery fire. What were the men to do ? Turn to their officers for orders ? Out of the twenty-five officers who had left our trench little more than an hour before, only one was now unhit. To stay where they were, outflanked on both sides, was impossible, and the poor remnant of that gallant Battalion was forced to make its way back to the line of German trenches from which it had started. There they remained for the rest of the day, subjected to heavy shell-fire. Lieut. W. K. Tillie, the Battalion M.G.O., the only officer who was not a casualty, brought the Battalion out of the line the same night, when it was relieved and moved back to bivouac in a field near Noeux-les-Mines.

Had the Battalion had more experience of modern warfare, it is probable that a nucleus of officers, N.C.O.s, and specialists would have been left behind at the transport lines before entering a big battle ; but, as it was, every available officer and man went into the line, and with casualties to every combatant officer except one, and over 550 other ranks out of about 900, the organisation of the Battalion was temporarily ruined, and what remained of one of the finest and most gallant battalions that ever left England required months of careful rebuilding before it was a semblance of its old self. To those who had seen the Battalion grow from its infancy and who lived to look back on that day, there remained, and will remain, an everlasting regret that it should have been frittered away in two short hours on a hopeless task; but at least we know and are proud that the task was undertaken with a determination and bravery which could not have been excelled, and the fact that it was impossible is the only reason why it was not accomplished.

Of the twenty-four officer casualties, thirteen were killed. Major Brock-Hollinshead, the 2nd-in-Command, Lieuts. Gibbs and Lawson of A Company, Lieuts. Smith, Bigsby, and Don of B Company, Capt. Edwards and Lieuts. Watts and Harris of C Company, and Capt. Robertson-Ross and Lieuts. Burrell, Beer, and Plant of D Company; Col. Vansittart and Major Chillingworth, Capts. Middleton and Hutchinson, Lieuts. Old and Ell were badly wounded, and were picked up by the enemy after lying exposed in No Man's Land for many hours; and the only five to reach England were Major Pullman, Capt. Grant, and Lieuts. Green, Lewis, and Hough.

For gallantry in this battle Major Pullman was awarded the D.S.O., Lieut. Tillie the Military Cross, and Col. Vansittart and 2nd Lieut. Don were mentioned in dispatches. After the Armistice a further award of the Distinguished Service Order to Col. Vansittart was announced, and was a most fitting tribute to his gallant personal leading and a great compliment to the

Battalion which he had brought to such a high state of efficiency during the period of training.

When he was back in England in 1917, Col. Vansittart closed his report to the War Office with the following paragraph: "Might I add that the 8th Royal West Kents actually got from half to three-quarters of a mile east of Hulluch—that although the fire was very heavy and from every direction, officers and men alike showed not the least semblance of any check or stop. They advanced as if on parade and under perfect discipline till they reached the enemy's undamaged barbed-wire entanglement, beyond which they were unable to go; and here our losses were very heavy."

Writing to Mrs. Vansittart after the battle, Gen. Mitford said: "Nothing too much can be said for the 8th Royal West Kent Regiment; notwithstanding the shell and machine-gun fire under which they advanced over open slopes destitute of cover, they never wavered an instant, but moved on towards their objective as if on parade."

The other three battalions of the Brigade were nearly as badly hit, and the total casualties were approximately 78 officers, and 2,000 other ranks, out of about 3,600.

The following address by Brig.-Gen. B. R. Mitford, C.B., D.S.O., was read to the 72nd Infantry Brigade at the Parade Services on October 3, 1915:

"72nd Infantry Brigade,

"Last Sunday the Brigade went into action for the first time, only a year after they came forward at their Country's call.

"The way the Brigade advanced under very heavy machine-gun and artillery fire from flanks and rear has evoked the approbation of the Divisional and Corps Commanders. You were an example in steadiness and determination to carry out your task, not only to the new Armies, but also to seasoned troops, who could not have done better than you have.

"As I say, you carried out your task, but you had to retire. Yet do not think it was a failure, for it was not, as you caused 16 battalions of the enemy reserves to be brought up into your area and taken away from the French just South of us, thereby enabling the French to make an appreciable advance.

"I should like all of you who know the relatives of those who are no longer with us to tell them how gallantly they fought and how nobly they served their Country, in whose service they fell, and what prestige they brought on the names of the Regiments to which they belonged.

"Men of the Queen's, Buffs, East Surreys, Royal West Kents, you have added glory to the ancient Regiments of which you are the children. You have made for the 72nd Infantry Brigade a name which none of you can be anything but proud of, and which I know, in the future, you will never allow to diminish.

"I feel it a great honour to have had the chance of commanding such troops on service, and I shall never forget the ground about Hulluch village."

PHASE II

TRENCH WARFARE IN THE YPRES SALIENT AND AT MESSINES

September 1915—*July* 1916

AFTER Loos the work of rebuilding the Battalion was commenced at once.

From the plains of Hulluch the remnant of the Battalion was withdrawn to Mollinghem, thence to Houtkerque, and finally to the muddy camps at Reninghelst.

Here, on Oct. 6, Maj. J. C. Parker came to take over command of the Battalion. Large drafts of officers and other ranks had been arriving from the 3rd and 9th Battalions, but it was not possible to give many days to the important work of reconstruction, and on October 15 Major Parker took the Battalion into the trenches at Spoil Bank, on the Ypres–Commines canal. Six-day trench tours with four-day intervals spent in the Reninghelst mud were the order of the day.

During the tour on the Bluff, November 3 to 7, the Germans blew a crater in front of our lines. This was distinctly an " experience " for the Battalion in the trenches for almost the first time. This trench work was a very trying time indeed. The state of our positions at St. Eloi,

Photo by] [*J. Weston & Son, Folkestone.*

LIEUT.-COL. J. C. PARKER.

Commanding the Battalion October 1915–August 1916 and January–October 1917.

where we held the trenches from November 12 to 19, was almost indescribable. Our trenches were towards the bottom of the slope east of St. Eloi village, and all water drained into the bottom of the valley. Rain was continuous, there were practically no shelters whatever in the front line, and the trenches were nearly waist deep in water. The enemy whizz-banged our tottering parapets at all hours, and the sniping from the trenches on the high ground was wonderfully accurate. The removal of wounded was a matter of extraordinary difficulty and danger. The rests at Reninghelst between trench tours were rests only in name. Large working parties had to go up every day to St. Eloi, the Bluff, or Spoil Bank. Parades were usually at 5.30 a.m., and the parties were conveyed to Café Belge in G.S. wagons. From there a march of two hours, often in a deluge of rain, brought the party to its work. About 5 p.m. the return journey from Café Belge would commence, and the tired party—usually depleted by a few casualties—would be swallowed into the mud of the camp again between 6 and 7. Training was almost impossible. It was with great relief that the Battalion heard during that terrible tour in the St. Eloi trenches that it was to be the last for the time being. On November 19 the Battalion came out of the trenches at St. Eloi, and marched, via Eecke, Ochtezeele, and Houles, to Bonningues, a small village some ten miles from St. Omer.

Towards the end of this march Capt. N. I. Whitty, 2nd Battalion, joined as 2nd-in-Command.

The Battalion arrived at Bonningues on November 25, 1915, and stayed there till January 4, 1916. Few who were with the Battalion during those five weeks will ever forget the stay there. The villagers were very kind to us, and the rest and peace after the terrible shock of the Battle of Loos, and the very trying trench life which had followed, were just what were required to knit together the Battalion and set it firmly on its feet again. New officers got to know their men, and the men got to know their platoon commanders. Training and refitting occupied most of each day, but on half-holidays and on Sundays one could roam on to the hills and gloat over the really beautiful scenery. The weather was fine on most days. During the month a good deal of football was played. In the Inter-company Competition B Company won, while in the Officers *v.* Sergeants' match the Sergeants lost by 14 goals to nil! B Company also reached the semi-final in the Divisional Inter-company Competition. An inter-company cross-country run was arranged, forty runners per company, and was won by A Company.

The whole Battalion spent a very happy Christmas at Bonningues.

Before we watch the Battalion marching into the line again, we might recall the names of some of the officers and N.C.O.s of the Battalion at this time. Lieut.-Col. J. C. Parker was commanding the Battalion, with Capt. N. I. Whitty as 2nd-in-Command. Capt. W. K. Tillie, M.C., was adjutant. The company commanders were: A Company, Capt. W. Wood (one of the family of Wood which has been so long associated with the Royal West Kent Regiment); B Company was commanded by Capt. W. V. Ross (who had till recently been an N.C.O. in the Argyll and Sutherland Highlanders); C Company, Capt. T. P. P. Walker (from the 9th Battalion); and D Company, Capt. C. F. Penton (who had been with the 8th Battalion in England). Lieut. Evans was Quartermaster. The Warrant Officers were: Regt.-Sergt.-Maj. Lee, Regt.-Quartermaster-Sergt. Dadd; A Company, Coy.-Sergt.-Maj. Elson; B Company, Coy.-Sergt.-Maj. Ackhurst; C Company, Coy.-Sergt.-Maj. Verrall; D Company, Coy.-Sergt.-Maj. Acland; Cpl. W. C. Carr was chief clerk in the orderly room, and Sergt. Wood was Transport Sergeant.

During the rest at Bonningues two very successful Sergeants' suppers and concerts were organised under Regt.-Sergt.-Maj. Lee.

A few days before leaving the rest area the whole of the 72nd Brigade, still under the command of Brig.-Gen. B. R. Mitford, C.B., D.S.O., were reviewed by Gen. Plumer, whose army in the Ypres Salient the 24th Division was about to join.

The usual reconnoitring party set off early on the morning of December 28, by motor bus, to reconnoitre the Boesinghe Area with a view to taking over those trenches. On arrival it was found that a gas attack had been launched by the Boche a few days previously, and he had attempted to come over, but had been driven back by the 47th Division, whom he had then shelled mercilessly and with little reply for thirty-six hours. The 47th Division had suffered many casualties and were very weak in numbers, one company being down to twenty-three men. The reconnoitring party was hailed with joy by the men in the line. The trenches were in a bad state and reeking of asphyxiating and lachrymatory gas.

Having spent thirty-six hours in the trenches and thoroughly reconnoitred the area and the Canal Bank Dug-outs, the party returned, not very favourably impressed with what it imagined to be its next trench area.

The next few days were spent by the Battalion in rigorous gas training, a special expert being sent down by General Headquarters to lecture on the subject. On January 2 orders were received that the Battalion would entrain on the 4th and 5th, and proceed to Poperinghe, preparatory to taking over the Hooge Sector. Although not very sweet on the Boesinghe Area, the name of Hooge struck an almost more unpleasant chord with some of us.

On January 5 the Battalion, now quite up to strength again, entrained at Audricq for Poperinghe.

On January 6 the Battalion moved up and took over from the 12th Manchester Regiment the right sector of the Hooge Trenches B1–C3, being disposed as follows: B Company on the right, A Company on the left, D Company in support, C Company in reserve.

A quiet five days were passed, B Company being slightly shelled from the direction of Hill 60, and suffering a few casualties.

On the night of January 11 the Battalion moved back to Belgian Château, and remained there in support till the evening of January 14, whence it moved back to Ouderdon, forming part of the 24th Divisional Reserve.

Here from January 15 to 26 training was continued.

On January 27 the Battalion took over the left of the left sector of the Hooge trenches from the 8th Queens, where it remained till January 31. A very quiet few days were spent in the trenches, only two casualties being recorded.

On the night of January 31, B Company and half C Company moved back to camp at Ouderdon, and A Company, D Company, and half C Company moved across to the left in support to the Railway Wood Trenches, A Company and half C Company being in the Ramparts, and D Company in direct support.

Working parties were the order every night while in this position, assisting the Mining Company in making deep dug-outs in Hill 20 and Leicester Alley.

On the night of February 5, A and D Companies moved up and took over trenches A1, A2, and Hill 19 and 20 from the 8th Queens, D Company taking over A1 and A2. A1 was a trench entirely unconnected with the remainder of the system, there being no access to it except a duckboard walk " over the top "

As this was about the date of the Kaiser's birthday, the Boche became very bombastic, and prophesied large attacks, but nothing of any importance took place.

On the night of February 8 the Battalion was relieved in this sector by the 2nd Leinsters, and moved back to rest in Camp C, where it remained for a week, training being continued, chiefly in bombing.

On the 16th the Battalion moved up and took over the Belgian Château Dug-outs from the 8th Buffs.

The Boche was now getting considerably more aggressive all round the Salient, and it was thought that he might attack during the coming spring.

On the afternoon of the 19th a disaster occurred which shook the Battalion very considerably, and carried off some of its best officers. The Battalion was due to move into the Sanctuary Wood Trenches about 5 p.m., and all orders had been issued to that effect. At about 3 p.m. a long-range naval shell fell in the neighbourhood of B Company's Headquarters Dug-out, followed by another at a short interval which obtained a direct hit on the dug-out. Capt. Ross and Lieut. Atkinson, who were in the dug-out, were badly wounded. Capt. Ross subsequently died, and Lieut. Atkinson lost a leg. As D Company's Headquarters' Dug-out was close by, the Officer Commanding D Company at once left his dug-out and tried to summon aid and render what assistance he could. He had only got about six yards from the entrance of D Company's Dug-out when another shell followed, which burst right in the dug-out entrance, severely wounding its three occupants, all of whom subsequently died of their wounds. Thus perished Lieut. W. L. Wigan, a regular officer, who had served with the 1st Battalion, and who was universally loved by all ranks. His loss to D Company could not be adequately expressed. Also 2nd Lieut. Cadell of D Company, a bright, cheery boy, and 2nd Lieut. G. M. Stanton, who had only been with the Company about a week.

The commotion caused by this unhappy event seems to have been observed somehow by the Boche, who shelled the whole length of the dug-outs for the next hour or so, though without inflicting much further damage.

D Company was thus left to take over its portion of the Sanctuary Wood Trenches, B1 and B2, with only Capt. Penton of commissioned rank. Lieut. S. B. Smith was therefore ordered to report to D Company for this tour in the trenches. About 5.30 p.m. the Battalion moved up and relieved the 8th Queens in the right of the Right Sector of the Sanctuary Wood Trenches.

During this tour, February 19–23, considerable apprehension of attack would appear to have existed in high quarters. On the 21st the following messages were received in the front line in the space of three hours:

(1) "Gas reported by troops in Boesinghe Area, coming from the north-east."

(2) "Aeroplane reports enemy massing behind Sanctuary Wood."

(3) "Prisoner taken at St. Eloi this morning states that an attack on Sanctuary Wood is imminent."

About two hours later another message came through:

"Cancel statement of prisoner taken this morning at St. Eloi; prisoner misunderstood."

Nothing, however, came of all the reported Boche preparations, and the Battalion moved back to Belgian Château Dug-outs on the 23rd after a quiet tour, during which a considerable amount of good wire had been put up to cover the front-line system.

About this date the Boche was becoming rather troublesome in and about the Salient, exploding a mine at Hooge and attempting to take the crater. Though causing a number of casualties in the 9th Sussex, he was completely stopped, and vented his anger in considerable shelling, destroying a number of the communication trenches behind the front-line system. He also persistently shelled the Belgian Château Dug-outs, though whether this was directed at the large number of batteries stationed about that spot or at the dug-outs is open to doubt. During the period February 24–27 spent in these Dug-outs he caused a number of casualties in the Battalion by shelling, D Company being particularly unlucky, as he managed one night to drop a shell into the dug-

SANCTUARY WOOD.

CAFE BELGE.
On the Ypres-Dickebusch Road.

out occupied by No. 13 Platoon, which killed four and wounded fourteen.

About this date also the Germans attacked and captured The Bluff from the division which occupied the trenches on the right of the 24th Division.

This success of the Germans caused considerable uneasiness at 2nd Army Headquarters, and it was thought probable that the 24th Division would be ordered to retake the Bluff; this operation was, however, carried out by the 3rd Division, who had just moved into rest as Corps Reserve. The taking of the Bluff by the Germans caused "a certain liveliness," however, in the 24th Division, and the system of reliefs had to be altered, owing to a whole brigade being kept in Divisional Reserve complete, instead of one battalion of each brigade.

On the night of February 27 the Battalion moved up again, and took over the Sanctuary Wood Trenches from the 8th Queens, the sector taken over being the right of the Right Sector and Maple Copse.

On February 29 orders were received to move back to Belgian Château, and the 8th Queens took over the Sanctuary Wood Trenches.

It was thought at first that the Battalion was going to take part in retaking the Bluff, but nothing transpired, and the Battalion remained at Belgian Château Dug-outs till March 3, when they moved up and took over the sector occupied by the 8th Queens. Coy.-Sergt.-Maj. Acland was wounded during the stay at Belgian Château.

This proved a quiet tour from March 3–7, though a considerable amount of work was put in building up the trenches, orders having been received that everything possible was to be done to strengthen the system, as it was believed that the Germans meditated an attack in the spring.

During this period the Battalion lost a valued officer in 2nd Lieut. Hamilton of C Company, an original officer of the Battalion from its inception. He was sniped while going down a communication trench to Company Headquarters.

A few casualties occurred during this tour, chiefly from rifle grenades, which the Boche had now begun to use extensively.

On March 8 the Battalion moved back into rest and took over A Camp at Ouderdon from the 13th Battalion Middlesex Regiment.

Here the Battalion was strengthened by the arrival of a fresh batch of officers, among whom was Sir Henry Hawley, an original member of the Battalion and afterwards Officer Commanding C Company.

During the period spent at A Camp, March 9–15, the Battalion had to find working parties, two companies strong, every night for work in the Salient, building machine-gun emplacements, and wiring and repairing trenches. These parties proceeded by train to Ypres, and then marched to their rendezvous. A number of casualties by shell-fire were caused on these working parties, the Germans showing the most uncanny knowledge of the whereabouts of the sites of proposed machine-gun emplacements.

On the night of March 15 the Battalion moved forward and took over the dug-outs at Belgian Château. Rumours of relief by the 3rd Canadian Division were now rife.

Working parties on the support defences south of the Lille Gate continued unceasingly ; much of this work was observed by the Germans daily, by the aid of their Fokker aeroplane, which at this time was superior to any British machine in use, and it was no rare sight to see two or more of our machines crashed daily by Fokkers.

The night March 20/21 saw the relief of the Battalion by the 4th Canadian Mounted Rifles, and the Battalion moved back to Camp F. It may safely be said that all ranks were glad to leave the Ypres Salient after nearly three months of monotonous trench warfare. Moreover, the Boche was undoubtedly meditating mischief, which culminated in his lightning attack on the Canadian Corps, an attack which the Canadians heroically withstood, though it cost them enormous casualties.

On March 22 the Battalion moved to Berthen, near Bailleul. Excellent billets were found for all, and the contrast of unshelled fields, spring weather, and peaceful nights soon put all ranks on good terms with themselves.

Training was carried out for three days, the usual reconnoitring party proceeding to the Wulverghem Sector on the 24th to reconnoitre the new area to be taken over from the Canadians.

The new sector was found to be extraordinarily quiet, and the lack of shelling amazing, whole houses and farms being untouched by shell-fire within 1,400 yards of the front line.

On the 26th the Battalion moved to Dranoutre and took over the rest area for the night, the men being well pleased with the hutments and the look of Dranoutre, which at this time was absolutely untouched and full of its original inhabitants, plying a thriving trade with "Tommy."

On the 27th the Battalion moved up to Wulverghem and took over trenches C3 and C4, B1 and B2, facing Messines, from the 1st Canadians, moving in by sections in daylight about 4 p.m., the relief being accomplished without loss.

Headquarters took over St. Quentin Cabaret, a small inn within 1,000 yards of the front line, absolutely intact and undoubtedly the best headquarters in the whole of the British front at that time. The support company lived even nearer to the line in Souvenir Farm, also intact.

During the months previous to our taking over there had been something like an understanding between our artillery and the enemy's: "If you don't shell, we won't." So estaminets and small shops were open in and around Wulverghem, less than three-quarters of a mile from the front line, and farmers were working in the fields just as close.

There was only one line of trenches on our front, and, as was the custom in those days, every firebay had a section of men in it. About 400 yards behind the line were absurd Strong Points built overground of sandbags. Each was held by a platoon. The two on our front were known as S.P.4 and S.P.5. The chief incident during our first five-day tour in the front line was the destruction of S.P.5 by enemy artillery fire. He shelled it very accurately for over an hour with 5·9s. The garrison stuck it out under Sergt. "Miffy" Mills, who, though badly shaken, was merry and bright when it was over. Cpl. Hogg was wounded.

After five days we moved into support, not very far back, all living in farm buildings. Battalion Headquarters and A Company went to Tea Farm, B and C Companies to Aircraft Farm, and D Company to Kandahar Farm.

Trouble began during this tour in support. Our artillery had given up the policy of peace, and had been steadily strafing the Hun during the previous days. So *he* started. First of all on Tea Farm. At 5.30 a.m., on April 6, he put a 5·9 shell into the loft of one of the barns where the Battalion bombers were sleeping. It was real bad luck that he hit the place first shot, because we had trenches a short way off to go to; and before the next shell came over a few minutes after, everyone not actually employed on attending to casualties was safely away. That first shell accounted for thirteen casualties, eight killed and five wounded. He shelled the farm then for two hours, and off and on all day for three days, until it was a wreck. Cpl. Bates greatly distinguished himself evacuating the wounded under heavy fire. Battalion Headquarters moved up to Kandahar Farm with D

Company, and we had peace for the rest of the five days out.

Our next tour in the trenches was uneventful. We had a few casualties, including 2nd Lieut. Eldridge, who was wounded in the knee. On April 10 we were relieved by the 8th Queens, and moved back into reserve at Dranoutre, where everyone was fairly comfortable.

So things went on for a time—five days in the line, five days in support; five days in the line, five days in reserve.

At the end of April the enemy released gas against us. The 8th Queens and 1st North Staffordshire Regiment were in the line. We were in support and the 9th East Surreys were in reserve. We had to take up our battle positions, and assist in the work of getting up new gas-masks and rations after the attack was over. We escaped lightly, but the battalions in the line had fairly heavy casualties. Shortly after this it was decided to organise a big raid in retaliation for the gas attack, and our Battalion was ordered to carry it out. Maj. N. I. Whitty organised the raid, Capt. W. Wood was in command, and other officers were Lieut. N. B. Green, 2nd Lieuts. S. E. Dove, S. B. Smith, and T. Bowling. During the greater part of May the party were out of the line, training at Dranoutre. The sector had become decidedly " lively " by this time, and the Boche had " set about " most of the open farms in the district. The civilians had given it best and left.

On May 6, Capt. W. K. Tillie left to command

the 72nd M.G.C., and Lieut. S. J. F. Mills became adjutant.

On May 31 occurred one of the tragedies in the history of the Battalion. Capt. W. Wood was killed by a German sniper when out in No Man's Land patrolling with some of his raiding party.

It is impossible to exaggerate the worth of Capt. Wood's work while he was with the Battalion, or the sense of personal loss felt by all ranks when the news of his death reached us. Brave as a lion himself, he was a strict disciplinarian. His work and influence as Officer Commanding A Company had been marvellous, and the effect of his example upon the whole Battalion was seen at every turn. Perhaps one can sum up his record with the Battalion by saying that at first most of us were afraid of him, then we came to respect him, and last we came to love him. He was buried at Dranoutre on June 1. Gen. Mitford and his staff were present, and practically all the Battalion. Gen. Capper, the Divisional Commander, in a letter to Col. Parker, spoke of the personal loss he had sustained by Wood's death, and alluded to him as the finest company commander in the Division. " Quo fas et gloria ducunt."

On the night of June 4/5 the raid was attempted. 2nd Lieut. S. B. Smith behaved with great gallantry in trying to cut the enemy's wire, but it was found impossible to get through. We had a number of casualties, and 2nd Lieut. Bowling was killed. 2nd Lieut. Smith, although he had been working hard nearly all night and was dead beat, went out just after dawn and got in Bowling's body.

June 1916 will be remembered long by all who were with the Battalion. We were out of the line, in support again, for the enemy's second gas attack on June 17. The Queens and the North Staffords had the misfortune to be again in the line. This time the enemy used phosgene for the first time, and the Brigade suffered severely. Most of our Battalion were in the front line or in the communication trenches as working parties, and they behaved magnificently. All night and long into the next day they helped with casualties. Gas discipline among our men was then, as it always has been, very good, and we had comparatively few casualties. Apart from these few days in support, which were far from a rest, we were in the front line the whole month, and after the enemy gas attack it was no picnic.

It was arranged that we should release gas in retaliation, and all labour was devoted to installing the cylinders in our trenches. Work was carried on at feverish speed, and at the same time the enemy had brought up his well-known "travelling circus" of trench mortars opposite us, and he made a practice of blowing us up every evening. The trenches came to be known humorously as the "Wulverghem Empire"—Two Houses Nightly.

On the 28th we released our gas, and the Queens followed it up and raided the enemy's trenches. The wire had been previously cut by artillery, the raiding party got in successfully and captured about six prisoners. In this raid one of our officers, 2nd Lieut. Reynolds, who had been transferred to the Queens, did splendidly. This night's work annoyed the Hun no end, and he made our positions a positive inferno. The next evening he set about us with trench mortars of the heaviest type, and fairly wrecked the trenches. All ranks behaved with the greatest gallantry, and were ready to repel an enemy attack at any moment. We had forty casualties, and 2nd Lieut. E. F. Vinicombe, of D Company, was killed.

On the next day, June 30, we were relieved by the 7th Australian Battalion, and moved out to Kortipyp Huts, very tired but greatly exhilarated by the experiences of the past weeks. We knew very little about the coming Somme offensive, but we had rumours, and we assumed, correctly, that we were " for it " there sooner or later. It was with surprise, therefore, that we received orders after a few days out at Meteren to go into the line again at Ploegsteert. Here some of us were served out with Australian hats, which we were to wear, and show ourselves in order to give the enemy the idea that the Australians were still in the line. Our stay in these trenches only lasted two days, when we were relieved by a battalion of the 20th Division, and moved out to Berthen. Here we remained for a few days until orders were received to move by train to Amiens, behind the Somme Battle. Then we knew.

The end of June 1916 saw the Battalion in good spirits, united and keen. We had had real hard trench experience, and the Battalion had proved itself worthy of the trusts which had been put upon it. All ranks looked forward with confidence, even with eagerness, to the Somme, and all the possibilities of getting at the Hun in the open and really finishing the war which it promised. How things actually turned out on the Somme are well known, and the next phase of the Battalion's history will give details of our part in that great struggle.

PHASE III

THE BATTLE OF THE SOMME

August—September 1916

After a few days at Berthen the Battalion entrained at Bailleul and detrained at Longeau, near Amiens.

By this time we had all formed a pretty clear idea of what was happening in the Somme Battle. News of the terrific fighting had been digested. We had read with pride of the work of our 7th Battalion in Trones Wood, where we were ourselves very soon destined to be, and we knew we were in for serious business.

During July 25 we marched seventeen miles to the village of Saisseval. It was intensely hot and very dusty, but we arrived at Saisseval about noon. This village was without exception the dirtiest and nastiest the Battalion had ever been quartered in. However, there was some excellent country round the village, and for a week we trained strenuously in the new methods of following a barrage and attacking in waves.

On July 31 and August 1 we moved to the Sandpits, near Meaulte, just outside Albert. We were here for ten days continuing the training. On August 8 A and B Companies were suddenly ordered up into Montauban. An enemy counter-attack was expected at Delville Wood, and it was apparently not certain whether the troops detailed for the Montauban defences would be able to get there. The two companies got into position in the afternoon, took over the dispositions, and arranged to continue work on the trenches at dawn the next day. However, during the evening the troops for Montauban arrived, and our companies marched back again to the Sandpits.

On August 9 orders were received to take over the front-line positions between Trones Wood and Guillemont the next day.

Early in the morning of August 10 our advance parties went up and thoroughly reconnoitred the positions. The rest of the Battalion followed in the afternoon. We were treated to quite a healthy bombardment on the way in, but the Battalion was finally settled in position before dusk. In front D Company were on the right, A Company on the left, in Edward Trench, B Company were in support just outside Trones Wood, and C Company and Battalion Headquarters in positions round the Sunken Road at the south-east corner of Trones Wood.

Such trenches as there were had only just been scratched out, and were mostly not more than a foot deep. From dusk till dawn work by the front companies was continuous. It was a case of digging for our lives, and at stand to in the morning every man had trench room nearly six feet deep. The positions were heavily shelled by the enemy in the morning, and our casualties would have been very heavy had it not been for the strenuous digging of the night before. As it was, very few were hit, and the experience was a very healthy object-lesson to all of us.

The unfortunate village of Guillemont in front of us was practically gone. Our shelling was continuous. During the day the place was usually obscured from view by clouds of red brick-dust from bursting shells, and at night the crashing of shells was continuous. The Hun did no persistent shelling, but he gave us hurricane bombardments at intervals, which were most unpleasant.

During the afternoon of August 12 a partial attack by the French on our right took place. We had a good view of the creeping barrage, and a very considerable amount of retaliation by the enemy on our trenches. Capt. Penton, Officer Commanding D Company, was wounded during the afternoon. In the evening things quietened somewhat, and we were relieved by the 1st North Staffordshire Regiment. We were out for forty-eight hours, just in front of the Briqueterie, during which time the whole Battalion was up at night digging a new trench, Nott Trench, some 400 yards in advance of our front line. On the 15th we went up again, D Company into Nott Trench, C Company into Edward Trench, B Company into the Trones Wood position, and A Company to Carnoy Craters. The enemy set about our new trench that night, and we had a bad time. 2nd Lieut. S. E. Dove, who was commanding D Company, was killed. There was no enemy attack, however. After two days we were relieved again, and the Battalion concentrated at Carnoy Craters.

It had been a stiff week. We had casualties to six officers and 146 other ranks. While in the line the Reserve Company had been working continuously in salving. As a consequence, huge dumps were made at Battalion Headquarters, which were used as forward dumps in future for all troops coming into the area.

An outstanding feature of this week's fighting was the magnificent work of Capt. McLarty, the Medical Officer. He was working almost continuously in the open in the Sunken Road by Battalion Headquarters, and he must have dressed hundreds of cases during the week. An abortive attack on Guillemont had taken place just before we took over, and wounded were coming down and being brought in by our stretcher-bearers continuously. Practically every shell-hole in front of us had a corpse or a wounded man in it. All night cries for " water " were heard. We had large parties out every night, and every soldier in front of us who was not past human assistance was got back.

On August 18 the 24th Division attacked Guillemont.

The 72nd Brigade was in support, concentrated at Carnoy Craters. The whole attack was on a frontage of fourteen miles. On our Division's front Guillemont Station and a post in the village near the famous Quarries were captured. In the evening we moved up, and as a first step we occupied our old positions in Nott and Edward Trenches. Battalion Headquarters were now in a dug-out in Edward Trench. A Company held Nott Trench, C and D Company Edward Trench, and B Company the Trones Wood positions.

On August 19 A Company was ordered up into the village to take over the post captured there and held by elements of the 73rd Brigade. From Nott Trench the Company marched in three sections into the village on a compass bearing after dark. The relief was completed before midnight, B Company having taken over Nott Trench. The Company in Guillemont was in a peculiar position. Our new line ran from the post in the village northwards to the Station. There were trenches of sorts here, and touch was established with the 3rd Rifle Brigade, who were holding the Station. But southwards and eastwards our post in the village was in touch with the enemy only, and that at very close quarters. The post itself consisted of shell-holes in a very early stage of consolidation. The quarries which faced our positions from the south at about 40 yards' distance were strongly held by the enemy, and he made the work of consolidation exceedingly difficult for our men by constantly harassing them with machine-gun and rifle-grenade fire.

By daylight on the 20th we had succeeded in getting a continuous trench facing south to the quarry, and two enormous shell-holes were consolidated as a defensive position facing east. The enemy seemed to have no settled position in the village, and he was evidently occupying shell-holes in a very irregular fashion. During the day he showed himself a great deal, and we did a lot of very successful shooting. In view of the enemy's unsettled position in the village, it was decided to make an attempt to continue our trench which faced south in an easterly direction, to form a jumping-off position for another attack on the village.

A large party of the 12th Sherwood Foresters (Pioneers) was to come up at night to do the digging, and the area where they were to work was to be previously cleared of and held from the enemy. At dusk, therefore, 2nd Lieut. A. Roscoe took out a big fighting patrol of a platoon in strength. He first worked eastwards from our shell-hole position to the eastern extremities of the village, then southwards towards the low ground in the centre of the village. It was a very difficult operation indeed, but his work was splendidly successful. The patrol had several encounters with isolated posts in shell-holes. One of these fights developed into a wrestling match in which one Hun was killed and one wounded. We had a few casualties, but succeeded in establishing two positions south of the site of the new trench. The garrisons of these posts remained out all night, frustrated all attempts by the Boche to filter northwards again, and covered the party digging the new trench. In the small hours of the morning of August 21 A Company was relieved by two companies of the 8th Queens, and moved back to Dublin

Trench, near the Briqueterie. It had been a highly exciting thirty hours up in the village, and the memory of it will live long in the minds of all who were there. The relieving feature had been the absence of shelling; our positions in the village were too near the enemy's, and the Boche positions east of ours were too indefinite for the enemy to shell us with safety to his own men. The other companies had been holding the old trenches, and had been subjected to the same old hurricane bombardments several times a day. We moved back to citadel Camp, near Fricourt, on August 22, and on the 25th to camps near Ville-sur-Ancre.

On the 27th the Battalion was inspected by Maj.-Gen. J. E. Capper, C.B., the Divisional Commander, who in a short speech thanked us for what had been done, and gave us his well-known doctrine of hatred of the Hun, as displayed by blood on the bayonet. He also spoke of our forthcoming work in Delville Wood. On this day company commanders had paid their first visit to Delville Wood, and Inner Trench, which had just been taken from the enemy in his final ejection from the wood, was reconnoitred. As a result of the reconnaissance it was clear to all of us that the enemy would never allow us to hold the trench without desperate efforts to retake it. The Battalion went into billets at Ribemont on the evening of the 27th, and stayed there till the morning of the 30th, when we marched to Fricourt. In the evening we went up to Delville Wood. Will anyone who was with the Battalion ever forget that night? At Bernafay Wood we were to be met by guides, and from there to Inner Trench the distance was about a mile "as the crow flies." The men were excessively heavily laden. Each company carried three days' rations, 25 petrol tins of water, 25 boxes of bombs, 12 boxes of flares, 12 boxes of Very lights, and some boxes of S.A.A. It had rained hard all day, and showed no signs of abating at night. The guides were picked up successfully at 9.30 p.m., and the crawl across Caterpillar Valley and through the wood began. Nominally there was a communication trench leading all the way, but there was little or nothing of it left, and the guides could scarcely be expected to keep direction. Delville Wood was notorious for changing its shape and appearance every hour. Details of that terrible night's march are innumerable, and cannot be dwelt on here. Most of them can be recalled by those who were there, and they can be imagined when it is stated that the rain was continuous, the shelling of the wood heavy and incessant, and the mud knee-deep at least. Added to this it should be remembered that the Battalion had already marched ten miles during the day, and the men were loaded up like desert camels. The leading company began to get into its position at 5 a.m. on August 31, but it was nearly 8 o'clock before the trenches were really taken over—eleven hours after we had picked up our guides a mile away!

A Company occupied Inner Trench and a new trench forward of the wood facing north-east.

DELVILLE WOOD.

THE MEMORIAL BOARD TO THOSE WHO FELL ON THE SOMME, 1916, PLACED IN GUILLEMONT CEMETERY.

B Company occupied Edge Trench in touch with the left flank of A Company. C Company were in support in the wood, and D Company in reserve behind the wood. Battalion Headquarters were in a dug-out at Longueval Church. Col. Parker had gone down sick with malaria and trench fever, and Maj. N. I. Whitty took over command of the Battalion. The Company Commanders were: A Company, Capt. H. J. Wenyon; B Company, 2nd Lieut. H. Flowers; C Company, Capt. Sir Henry Hawley, Bt.; D Company, 2nd Lieut. W. G. Lock.

We were allowed very little respite after the position had been taken over. If the night's march had been deadly, the day which followed was far worse. It was the day of the enemy's last and biggest attempt to retake the high ground from High Wood to Guillemont. We just had time to have our breakfast of sodden biscuits and cold bacon, when the shelling began, shortly after 8 a.m. It is not intended to paint lurid details of this shelling. Suffice it to say that Delville Wood went up in fire and smoke for five and a half hours. In Inner Trench was one deep dug-out. In this we kept five machine guns and crews of the 72nd M.G.C., also some of our Lewis guns and teams. The shelling round the dug-out was particularly heavy, but by the grace of God the enemy failed to demolish it. In Edge Trench cover was scanty and the shelling terrific. Most of the forward platoons of A Company were casualties before 10 a.m., and B Company on the left suffered very severely too.

The garrison of the two front companies became very thin, and two platoons of D Company from reserve were sent up to hold the centre.

At 1.30 p.m. the shelling suddenly and very obviously lifted on to the south edge of the wood. Shelters and dug-outs were immediately cleared, machine guns and Lewis guns were got into position, and our depleted garrison disposed to meet an attack.

Scouts crawled out in front to beyond the edge of the wood to watch a piece of dead ground there. On the right we were in fair order for meeting an attack. We had our machine guns and most of our Lewis guns intact, and quite a fair supply of ammunition. The enemy was some 600 yards away, and was bound to show himself clearly at 500 yards' range. On the left our position was more difficult, the enemy was much closer, and there was a covered way of approach opposite the front of the Battalion on our left. The men were tremendously eager. After terrific shelling which numbs the senses while it lasts, the reaction often took this form. As the first of the attacking troops began to trickle into view it was difficult to hold our rifle fire down. We kept perfectly silent, however, and, encouraged by this, the enemy came forward in considerable numbers till he arrived at an intermediate jumping-off trench some 500 yards away. Here he collected for half an hour, and word was passed down our line that as soon as he debouched from these positions in force, every rifle, machine gun, and

Lewis gun was to open fire. At about 2.30 p.m. the enemy came forward all along the front, at a slow trot. Immediately all along our line machine guns, Lewis guns, and rifles let fly at him. He was stopped dead within 50 yards of his trenches, while his men fell right and left. For a moment his line wavered, and the next his troops were running back in a mad rush to their original positions, followed all the way by our fire.

Baulked of his desire, the enemy shelled us again for an hour and an half, and then in the early evening he repeated his attempt. This time we knew exactly what to do, and we carried out the same programme as before with the same result. Inner Trench and Edge Trench were still entirely held by us at the end of the day. On the left of Edge Trench, however, the enemy had gained a footing in Orchard Trench, held by the battalion on our left, and B Company had a very nasty time in the evening. The Company Commander, 2nd Lieut. Flowers, was killed, and 2nd Lieut. E. G. Brown took over command. He and his Sergeant-Major, Coy.-Sergt.-Maj. R. Rankin, did splendid work, and organised their exposed flank for defence. Away on our right, too, the enemy gained a footing in the trenches of a battalion of the 7th Division, but he was ejected. A counter-attack the next day expelled the enemy from Orchard Trench and re-established the position all along the front. The enemy's great bid for the high ground had failed, and the night of August 31/September 1 was comparatively quiet.

Everyone was absolutely dead beat. In turns most of us managed to snatch some sleep that night, and the next morning we heard we were to be relieved during the day. The Queens relieved us in the afternoon, and we moved back into Check Trench in Caterpillar Valley, between Bernafay Wood and Delville Wood. The valley was stiff with our guns, and Check Trench was about thirty yards in front of their muzzles. At about 9 p.m., just when we were all hoping to get a bit of sleep, the enemy commenced a concentrated counter-battery bombardment with gas shell of the guns in Caterpillar Valley. We got the full benefit of this. Everyone was ordered to put on gas helmets (the old P.H.) at once. We spent a most uncomfortable night: the bombardment was kept up for five hours, and many of us had our first experience of sleeping in a gas helmet—quite possible, but most unpleasant. We had practically no casualties from that night's gas shelling.

We remained in Caterpillar Valley till September 5. The batteries were constantly shelled, and we usually got the benefit of it. On the first afternoon we had ten signallers buried in one firebay by an 8-inch shell. 2nd Lieut. Roscoe organised a digging party at once, and after two hours he succeeded in clearing the whole firebay; only one man was killed. A similar fate overtook another bay shortly after, and again all except one man were dug out alive. On the next day 2nd Lieut. Roscoe took up a party at daybreak to get rations up to the Queens, who were now in a new trench in front of Inner Trench. The party ran into very heavy shelling in Inner Trench, and had several casualties, including 2nd Lieut. Roscoe, who was severely wounded in the leg. He died of wounds at Corbie the next day. Throughout the fortnight at Guillemont and Delville Wood he had done magnificently. His cheerfulness was wonderful, and he was ready at all times for any job there was going. After 2nd Lieut. Roscoe was hit in Inner Trench, Sergt. Greenaway took the remains of the party on, though it was now broad daylight. They crawled across the open to the new trench, and delivered the rations and water successfully to the officers in command there.

On September 3 a big attack on Ginchy and Guillemont on our right took place, and on the evening of the 5th the 24th Division was relieved, and moved back to a camp near Fricourt. On September 7 the Battalion arrived at Vauchelles-les-Quesnoy, a delightful little village close to Abbeville, where we were very comfortable. A large draft of new officers joined us on our way, and the Battalion settled down to refit and train for what was to follow. As things turned out, we had finished with the Somme, and we were soon hurried off north to the 1st Army.

Our sojourn in the historic battlefields of the Somme had been crammed with incident. It coincided with the period in which the enemy was making his most strenuous efforts to drive us from the high ground, which gave the British their first sight of Bapaume. We had learnt what shelling meant at Ypres and Messines, and during the terrific bombardments at Guillemont and Delville Wood the Battalion was always steady. Apart from the holding and exploiting of the forward positions, the Battalion did a tremendous amount of work in their brief spells out of the front posts. As examples, one may mention Nott Trench, 800 yards long, dug in two nights right in front of our front line and under constant fire from the enemy, and the work of salvage done by the Reserve Companies in the Trones Wood area.

Of fighting the enemy we had had it at close quarters in Guillemont Village, and we had had it after a shattering bombardment in Delville Wood, where the enemy was not allowed to get within 400 yards of us. In both cases he had been beaten.

It is difficult to pick out any isolated cases of brave deeds among the countless acts of heroism of those weeks. 2nd Lieut. Roscoe's work has already been mentioned. 2nd Lieut. E. G. Brown and Coy.-Sergt.-Maj. R. Rankin in Edge Trench did wonderful work in an awful position in keeping the enemy at bay. The Battalion stretcher-bearers one and all did magnificently all through. They worked seemingly without sleep or food, and without ceasing, as also did Capt. McLarty, the Medical Officer. An act of two private soldiers may be mentioned as indicating very truly the spirit of the Battalion during the day of the enemy's counter-attack at

Delville Wood. Towards the end of the bombardment, when one front company had lost over a third of its men, the Company Commander considered it imperative to send back a report to Battalion Headquarters. He hesitated to do so for a time, because it seemed certain death to the men taking the message. Finally volunteers were called for, and two wounded men, Pte. Skerry and Pte. Tyrrell, came forward. The message was given to them, and they were told to deliver it at Battalion Headquarters, and then to go down to the field ambulance. The wood was a perfect inferno at the time, but they started off with the greatest bravery, although one of them could scarcely walk. Something more than an hour later the Company Commander was astounded to see them back in Inner Trench, and assumed they had not been able to get through the barrage. The fact was, however, that they had not only got back to Battalion Headquarters with the message, but they had insisted on bringing an answer back to their Company Commander rather than go straight back to the safety of the field ambulance. As the result of this message the reinforcements from D Company already mentioned came up to fill the gaps in the two front companies.

One other case may be mentioned. At the end of a long night's work in front of Guillemont, dawn broke with a slight mist. Cries for "water" could still be heard in front, so under cover of the mist a large party went out and brought back many wounded men. While they were out, a cry was heard from a long way in front. An officer who was with the party went forward as far as possible and discovered that the wounded man was in the enemy wire. The stretcher-bearer party was following up, but the officer ordered them not to go forward, as it seemed obviously hopeless to get the man back. The whole party therefore returned to Edward Trench. It was then discovered that one man, Pte. Allchin, was missing. A few men went out again to find him, and after about an hour he was seen coming back from the enemy wire carrying the wounded man on his back. That he had got there and back was nothing short of a miracle, because the enemy was exceedingly vigilant at this time, and he was sniping incessantly.

The fighting on the Somme had taken a heavy toll of the Battalion. The Battalion memorial erected in Guillemont Cemetery gives the names of 3 officers and 91 N.C.O.s and men who were killed or who died of wounds. Our total casualties were 13 officers and about 300 N.C.O.s and men.

The impression left in the minds of most people who fought on the Somme was that the enemy was being taught there once for all that the British Army was going to beat the Germans at all cost.

The 8th Battalion of the Royal West Kent Regiment can justly claim to have done much to drive home that lesson to the enemy at Guillemont and Delville Wood.

PHASE IV

TRENCH WARFARE AT VIMY RIDGE, HULLUCH, AND MAROC—THE ADVANCE ON LENS

September 1916—*April* 1917

Our stay at Vauchelles was pleasant, but short. The billets were all quite good, and Battalion Headquarters occupied sumptuous quarters at the château of the Baron de Rocquigny. Reorganisation was taken in hand.

The commands were as follows:

Commanding Officer, Maj. N. I. Whitty; A/Adjutant, Lieut. H. S. Brown, vice Lieut. and Adjutant S. J. F. Mills, who broke his ankle as the result of a fall from his horse.

Company Commanders: A Company, Capt. H. J. Wenyon; B Company, 2nd Lieut. E. G. Brown; C Company, Lieut. Sir H. C. W. Hawley; D Company, 2nd Lieut. W. G. Lock.

On September 19 we entrained at Abbeville under orders to move to the 1st Army Area. After a four days' halt at Sains-les-Pernes, where training and reorganisation were continued strenuously, a further move was made to Haillicourt, near Bruay, for a couple of days before the Battalion arrived in the new area behind the Vimy Ridge and went into Divisional Reserve at Gouy-Servins on the 26th. The civilians were inclined to be hostile here, and we were not very comfortable. The Battalion was billeted in the château, which was in a very dilapidated condition. On October 2 we moved up to the front line in the right sub-section (Berthonval Section), and spent an uneventful spell of eight days there. The Hun did a certain amount of trench mortaring, but after the Somme these trenches were really a peaceful home from home. The communication trenches were in very fair condition, but the front line was not so good. A, C, and D Companies were in the front line, with B in support.

On October 10 the Battalion went back into support in Zouave Valley, Coliseum, and Alhambra Trenches and Cabaret Rouge. Here the chief sports were working parties of all sorts, and rat-hunting. On the 18th we moved to Gouy-Servins again, and while there the officers gave the sergeants a good trouncing at football, and later played a very even game with the whole of the rest of the Battalion.

Orders were soon received for us to move up to the Hulluch Sector. On the 24th the Battalion marched to Mazingarbe, and on the 25th moved into the line opposite Hulluch, relieving the 14th Highland Light Infantry, a bantam battalion of the 40th Division.

Many tragic memories were here revived, for we were holding the very trenches over which the attack on Hulluch had been launched on September 26, 1915.

The 72nd Brigade was on a two-battalion

Photo by] [*Swaine, London.*

LIEUT.-COL. N. I. WHITTY, D.S.O.

Commanding the Battalion August 1916—January 1917 and October—December 1917.

front, holding the left sector of the 24th Divisional front. We shared the right sub-sector with the 1st North Staffordshire Regiment. The trench system consisted of a front line running along the Lens–La Bassee road, a close support 40 yards in rear, a support line "Reserve Trench" 600 yards behind, and a reserve line "Tenth Avenue" for the Battalion in support about 1,000 yards behind "Reserve Trench." The Battalion in Divisional Reserve was in billets at Philosophe. The forward trenches were fairly good except on the left, but they had been recently knocked about considerably, and one portion of the front line on the centre company front, which jutted out rather prominently into No Man's Land, was practically extinct. A trench called New Cut was dug across later on to replace this. There was another very bad place 200 or 300 yards in length between the centre and left companies which was unoccupied. The communication trenches on our front, Posen Alley, Broadway, and Vendin Alley, were very poor.

We found the Hun was very out of hand on this front. On our right company front No Man's Land was about 100 yards wide, and it increased to about 300 yards on the left company front. The enemy had a different treatment for each company. On the right, where the enemy held Posen Crater, a big chalk mound where a mine had been blown some time before, he had developed his light trench mortars, "Grenatenwerfer," or "Pineapples," to an extraordinary extent. On the centre company, where our trenches were bad, he used medium trench mortars, "Rum Jars," very extensively, particularly on the roof of Company Headquarters. The left company used to be shelled a good deal, and medium trench mortars dropped there quite frequently as well. We went into the line in the early morning of October 25, A Company on the right, B Company in the centre, D Company on the left, and C Company in support in Reserve Trench. Battalion Headquarters were in dug-outs in Curzon Street, 500 yards behind Reserve Trench. During the day all was quiet, but as soon as it got dusk in the evening the fun began. On the right the Hun fairly let fly with his "Pineapples," in the centre he began to drop "Rum Jars" with startling accuracy on the roof of Company Headquarters, and on the left he served up all kinds of stuff. We had very little means of retaliation, and something had to be done. By the second night we had arranged one or two artillery concentrations, which could be switched on to various parts of his trenches, and on the right we had got one or two rifle-grenade batteries in position for retaliation for his "Pineapples." On this night 2nd Lieut. F. G. Yeo, who had just joined A Company, was killed by a "Pineapple," and his platoon sergeant, Sergt. Champion, was badly wounded and died a few days after. For a fortnight a desperate duel for supremacy in these trenches went on. At first we could only manage to give him about two rifle grenades for each "Pineapple" he sent over, with a mild artillery strafe if things got very bad. But by the end of the fortnight he must have been getting twelve rifle grenades for every "Pineapple." Also, our artillery liaison had been greatly improved. Officers who were with the Battalion at that time will remember the code words, "Air," "Love," "Hate," "Fortnum," and "Mason," among those used to bring down fire and brimstone on the head of the Hun. The Stokes mortars were really being used as a mobile weapon. We made temporary emplacements all along the trench during the daytime with a few sandbags, and during the night 2nd Lieut. H. J. Dunn, with a party of men, used to carry a mortar along from emplacement to emplacement, firing off a couple of rounds as required here or there. The rifle-grenade batteries were manned all the time, but no firing took place until the friendly Boche saw fit to commence.

It was a most amusing fortnight. It would be difficult to say how many hundreds of rifle grenades and Stokes shells fell into the Hun trenches. One morning, when all was quiet, a company commander was looking at some old boxes of Mills bombs in the front line. In order to test them, he threw one over the parapet. It exploded all right—but the parties with the rifle-grenade batteries in the support line thought it was a "Pineapple" going off, and before you could say "knife" a hail of rifle grenades dropped into the unfortunate Posen Crater. By the end of the first week in November the Hun had given it best, and he was kept well in his place during the rest of our stay in this sector. We got on steadily with improving the trenches, and a lot of new wire was put up, and constant patrolling gave us the complete mastery of No Man's Land. Normal tours up here were six days in the line, six days in support in Tenth Avenue—with rat hunts and working parties galore to keep us busy; six days in the line again, and then six days in reserve at Philosophe. We remained in the sector for nearly four months, and it was on the whole a very happy winter.

During the greater part of this time A, B, and D Companies retained the right, centre, and left positions respectively, each one being relieved in turn by C Company when it was due for a tour in support. For the last two tours A and B Companies changed round in order to give B Company a rest from the centre sector, which had the reputation of being the most trench mortared. From the neighbourhood of Battalion Headquarters we had a very fine view of the whole of the enemy's trench system and the country behind his line, and sniping with artillery became one of the principal amusements. It was great fun chasing groups or single Huns about behind their lines with whizz-bangs or sniping with a 60-pounder at the intrepid cyclist who rode into Hulluch regularly every afternoon, and the effort to knock down Wingles Tower was another constant sport for the heavier artillery. We got to know the gunners very well in these days. In addition to having the liaison officer actually with the Battalion, most of their

PHILOSOPHE VILLAGE.
The billets of the Battalion in reserve.

POSEN CRATER.
In the enemy front line near Hulluch.

observation posts were in the neighbourhood of our headquarters, and the F.O.O. usually dropped in to see us on his way up to ask if anyone wanted to do a shoot.

We used to have good times at Philosophe for six days in twenty-four. It was just a one-street miners' village, with the outer shells of most of the houses intact, but lacking such comforts as window-panes and furniture. Our first tour the billets were far from comfortable, but we and the other battalions in the Brigade made improvements every time we were there, and when we left the sector it was a very satisfactory rest area. There were still some very dirty civilians about, mostly those connected with the mines, and some of them did quite a good trade selling odds and ends to the troops at a very large profit. Bethune was only a few miles away, and was very popular. Shops and estaminets were open there almost as in normal times, and divisional cinemas or concert parties often gave their shows in the town. We sometimes got motor buses or lorries to take a proportion of the Battalion into Bethune or to an entertainment elsewhere, and several very successful Battalion concerts were held at Philosophe during the winter.

Just before Christmas a draft of 200 arrived to make up our losses on the Somme. Christmas Day was not very exciting. We moved into the line from Philosophe on the morning of December 24, but we had all the excitement we wanted that day, as the Boches elected to give us one of the worst strafes we ever had in this sector.

Christmas was kept up during our next visit to Philosophe. Each company had one day set apart for it, on which the concert hall was at its disposal. The programme consisted of some form of sports or football in the morning, a colossal midday meal washed down with a plentiful supply of beer, and the afternoon divided up between the Divisional Band, a company concert, and a big tea. Everyone thoroughly enjoyed the Christmas of 1916, although it was celebrated in 1917.

From the beginning of January until a day or two before we left this sector, on February 12, the weather was very cold. There was some snow at the beginning of this long period, and the frost held so consistently that it was lying on the ground for over five weeks. All this time it was perfectly dry, and the trenches were in a far better state than at any other time during the winter. When we left the sector the thaw was just making itself felt, and we fear that our successors must have had a lot of trouble repairing the trenches, which were bound to collapse as the frost came out of them. During the latter days of January we were preparing a raid to be carried out by a party of C Company, under Lieut. H. L. Lewis and 2nd Lieut. K. Pfeuffar, but we were relieved before it could take place.

On January 20 Lieut.-Col. J. C. Parker returned after an absence of over four months, and resumed command of the Battalion.

Our last days at Hulluch were marked by raids and counter-raids, and a considerable increase in strafing all round. The 9th East Surreys captured a prisoner in a daylight raid on the left battalion front, and inflicted several other casualties on the enemy, but the 1st North Staffords had a bad time in attempting to enter the enemy trenches at night. The Hun was apparently sitting up waiting for them. The enemy made one or two attempts to raid our trenches in return without success, one of them being repulsed by the North Staffords, who captured half a dozen prisoners in addition to killing a number of the party.

On February 12 we were relieved by the 13th K.R.R.C. of the 37th Division, and moved out to rest at Busnettes, near Lillers, spending one night at Noeux-les-Mines on the way.

We were really sorry to leave Hulluch. We had been in the area so long that we knew its advantages and disadvantages, and were able to make full use of the former and avoid the latter to a great extent. Also we had done a lot of work to improve the accommodation, both in the line and the rest billets, and it always seemed to be our fate to leave to others an area which we had made comfortable, and move to a new front which had been neglected by our predecessors.

It had been a strenuous winter of trench warfare, but, knowing how unpleasant war can be at its worst, we must admit that it was a comparatively pleasant period and we had very few casualties.

The Battalion was very scattered at Busnettes, billets were not too good, and there was practically no ground at all available for training, which had to be carried out on the roads. However, it was a good rest, and we were more or less out of sound of the guns for a time. Lillers was quite close, and most of the Battalion paid one or two visits there. Battalion Headquarters was provided with a piano, and the officers will remember the wonderful playing of Lieut. H. J. Dunn, which helped to pass many pleasant hours.

At the beginning of March orders for a move were received, and at 9.30 a.m. on March 3 the Battalion left Busnettes and marched via Choques to Fouquereuil, where it was billeted for the night. The next morning the march was resumed via Drovin, Noeux-les-Mines, and Petit Sains to Bully-Grenay, where dinners were taken on arrival at 1 p.m., and where a few hours' rest were obtained in billets before moving up into the line at dusk. The part of the line being taken over by the Division was from immediately north of the Vimy Ridge, about Souchez to the Double Crassier, south of Loos. The 72nd Brigade was on the left of the Division in the Calonne Sector, with the right battalion holding the line east of Calonne, known as Calonne I, and the left battalion holding the trenches east of South Maroc, known as Calonne II. The support Battalion Headquarters and three companies were in cellars in Calonne ready to man the Calonne defences, and the 4th Company was in support to the left battalion in cellars in South Maroc. The reserve battalion was in billets in Bully-Grenay.

On the evening of the 4th we took over Calonne II from the 14th Canadian Battalion, A Company being on the right, with B Company in support in the southern portion of Maroc, near Fosse 11 de Bethune, and C Company on the left, with D Company in support. The trenches were in a very bad condition, particularly the main communication trenches, Edgware Road, Neuf Alley, and Treize Alley—so bad were they, indeed, that most people preferred to walk "over the top," even in broad daylight, to plodding painfully along the trench up to the knees in a particularly thick and clinging specimen of mud. The state of the trenches was not due to the fact that they were not revetted, but to neglect, as duck-boards and A frames were discovered after removing a couple of feet of mud, which had been gradually allowed to cover them. Needless to say, the Battalion set to work to improve this state of affairs, but it was some time before any great improvement was noticeable. We were indebted to the pioneer battalion, the 12th Sherwood Foresters, for one of the finest specimens of communication trench we ever had the pleasure to use. In a remarkably short space of time, considering the state when they started, they turned Edgware Road into a model, which for width, slope, and draining might have come straight out of a text-book.

Each of the two companies in the line held its trenches by means of three defended localities, with gaps of 200 to 300 yards between them filled with wire. Patrols used to move about between these localities at frequent intervals along the support line. There were a few deep dug-outs here and there in the support line, but the majority of the garrison had no cover either from the elements or the attentions of the Boche. On the right No Man's Land was only about 100 yards wide, and locality A got a considerable amount of light trench mortaring, but in the centre it widened to nearly 600 yards, and consequently the line there was not favoured with the attentions of the " Pineapple."

The most prominent feature on the Battalion front was the Double Crassier, situated on the extreme left, which consisted of two long parallel embankments of slag running out from Fosse 11 de Lens in the Boche lines. About 50 yards at the western end of each crassier, which was there about 50 feet high, was occupied by the British, C Company holding the southern crassier, and the northern being in the possession of the 6th Division, who were on our left. The situation on the crassier was somewhat peculiar. The enemy post was within easy bombing distance of our own, and consequently neither side cared about stirring up strife, as it was apt to become unpleasant for the instigator. Shelling was impossible, as the posts were too close together, but the garrison found these advantages were somewhat discounted by a feeling of insecurity and of being in the air, in two senses. Sniping on both sides was very deadly, and although it was most interesting to look from this point of vantage right back into the enemy trenches to the south,

THE DOUBLE CRASSIER.
In the front line at Maroc.

LOOS VILLAGE AND PUITS 15, KNOWN AS TOWER BRIDGE.

it was a pleasure which could only be taken with the assistance of a periscope.

Battalion Headquarters and the two support companies were in fairly comfortable cellars in Maroc, and were not often shelled, but machine-gun bullets used to whistle up the roads at intervals all night.

D Company supplied a platoon nightly to assist C in the defence of the crassier, and the rest of the two support companies were fully employed at work and carrying parties. The first tour passed without special incident, and on the night of March 8/9 the Battalion was relieved by the 1st Battalion North Staffordshire Regiment, Headquarters, A, B, and C Companies moving to Calonne, and D remaining in support to the North Staffords in Maroc. The cellars which served the Battalion as billets were fairly good, and many of them were provided with stoves, fuel being plentiful in the shape of beams from ruined houses. Baths were also obtainable, but these advantages were, of course, mingled with plenty of working parties. Considering its proximity to the line, shelling was very rare in Calonne, but, as at Maroc, machine-gun bullets were plentiful at night.

One of the principal forms of recreation, both in Maroc and Calonne, was rat-hunting. To assist in this excellent sport A Company Headquarters collected a pack consisting of Charles, who might, with a stretch of the imagination, be described as a fox terrier, and who had a perfect genius for catching rats, and John Willie, a large dog of no particular extraction, but perhaps with a touch of great dane about him. The rest of the party consisted of one person armed with a crowbar, usually 2nd Lieut. B. Wade, probably someone with a pick and a shovel, and various persons with revolvers and sticks. Wade would lead the pack to the scene of operations, and would mention the word " rat " to Charles, who would immediately take two sniffs, and then say in no uncertain fashion whether rats were present or not. Crowbar, spade, and shovel would then be employed until the victims were forced to break cover, when Charles and John Willie would do the rest. The sticks and revolvers were very seldom called into use. Thirty-seven was a record bag for a hunt, twenty-five being taken in one hour. At dusk great execution could be done with sticks or revolvers without the assistance of the pack.

Another diversion was the " Flying Pig." For the benefit of the uninitiated it should be said that the " Flying Pig " is the shell of a 9·45-inch trench mortar, is some two feet long, and has a tail with fins of about the same length. When it lands in moderately soft ground and explodes, it makes a crater some ten feet deep by twenty feet across, and is considered an unpleasant thing to have within fifty yards of one. The interesting part of a " Flying Pig " shoot consisted in counting the number of shells which fell in the Boche lines compared with the number which fell in our own. If the ratio was as much as four to one, it was considered a good shoot. Quite good sport could also be obtained by laying odds as to whether each shot would or would not be a dud. In the comparatively short time it was in action on our front it got two wonderful direct hits (which were not duds !) on our own front line trench.

On the night of the 14th/15th the Battalion relieved the North Staffords on its old front, Calonne II, B Company being on the right and D on the left. There had been a good deal of rain, and the trenches were, if anything, worse than on the previous tour. The next evening the enemy shelled the left company front for a short time, but did very little damage; the majority of the shells exploded very high in the air, and it is probable that they were a form of star shell, being used to range the guns for the raid which was attempted a few days later.

On the 18th, at 8 p.m., a shoot of old S.O.S. rockets was arranged to test their efficiency and give practice in their use. The Boche, though he used fireworks considerably himself, was always suspicious of other people playing with fire, and proceeded to shell our lines heavily. He even went so far as to put up his own S.O.S., and our artillery had to reply before he quietened down.

At about 4 o'clock the next morning, under cover of a heavy barrage, the enemy attempted to raid the Double Crassier, our part of which was then held by 2nd Lieut. F. Proctor and No. 13 Platoon of D Company. It is very largely due to the coolness and initiative of

L.-Cpl. Dent that the raid was repulsed without any casualties to us. At least four Boches, including an officer, were known to be killed on the Southern Crassier by us, and two or three more were killed on the Northern Crassier, where the Yorks and Lancs of the 6th Division had one officer and one other rank killed. Two heavy mobile charges were discovered with two dead Boches at the foot of the crassier, and it was evidently their intention to blow in the entrance to the tunnel in the Southern Crassier from which extensive underground workings started. The dead Boches gave us a very useful identification. L.-Cpl. Dent and Pte. Gale, who were with the support section, did very good work in repulsing the raiders with bombs when they had already worked their way round the forward section along the sides of the crassier. L.-Cpl. Dent was awarded a well-deserved Military Medal.

On the night of the 20th/21st, after some considerable enemy activity early in the evening, which was checked by our retaliation, we were relieved by the 1st North Staffords, and moved back to reserve in Bully-Grenay. The billets were moderately comfortable and shelling was rare. One afternoon, however, the enemy appeared to resent the presence of an Archie gun in the eastern part of the village, and dropped some 4·2's about the spot. Little damage was done, except for one shell which burst in the road not far from Brigade Headquarters and cut a water-pipe.

Very soon after our arrival in this area we

lost our old friend, Brig.-Gen. B. R. Mitford, C.B., D.S.O., who had commanded the Brigade since its formation in September 1914. Gen. Mitford left to command the 42nd Division, and was succeeded by Brig.-Gen. W. F. Sweny, D.S.O., Royal Fusiliers.

Even in reserve we found many working and carrying parties in store for us, the principal one being a company each night for wiring the reserve or "Village Line."

On the evening of the 26th the Battalion relieved the North Staffords in the line. This tour we took over in addition to our old front, the Northern Crassier, from the 6th Division. A Company were in their old front on the right, with B in support, while D Company, in the centre, and C Company, holding the crassiers, each found their own supporting platoons. During this tour there was a considerable increase in the enemy's trench-mortar and artillery activity. The whizz-banging of various portions of the line was fairly frequent and quite alarmingly accurate. The weather during the tour was very bad.

On the evening of April 1 we were again relieved by the North Staffords and proceeded to support, Headquarters and three companies in Calonne and one company in Maroc. As during the last tour in the line, there was a considerable increase in enemy artillery activity during our stay in Calonne. This was doubtless caused by the fact that preparations for the British attack on the Vimy Ridge and east of

Arras had been in progress since about the middle of March, and that the enemy was not quite sure how far north our attack would extend, and that some of the heavy guns employed in the preliminary bombardment were operating from our front.

On the night of the 7th/8th we returned to the line, B Company taking over the right, with A Company in support, and D and C Companies in the centre and on the left as before.

Throughout our stay at Maroc the patrolling of No Man's Land had been carried out regularly. The Boche was very seldom seen, and we had practically taken possession of the ground between the two trenches. During this tour patrolling became very active, two patrols going out nightly from each company in the line with orders to penetrate if possible into the enemy trenches and secure an identification. As was expected, however, it was found that the Boche wire was very thick, the whole line being protected by two and in places three good deep belts.

At dawn on April 9 the Battle of Arras opened with a terrific bombardment, but nothing could be seen of it, as there was a very heavy snowstorm at the time. Later in the day the weather cleared, and from the Battalion observation post in the school building the British barrage could be seen rolling across the more northern slopes of the Vimy Ridge. The weather continued bad for the next few days, alternate snow, hail, and rain making the trenches very unpleasant.

On the morning of the 12th the right brigade of the 24th Division, in conjunction with a Canadian attack further south, captured the Bois-en-Hache, and this apparently finally decided the enemy that the salient now formed west of Lens was an unhealthy place to stay in. That afternoon the enemy commenced to shell our trenches, and more particularly Maroc, in a way which we had never before experienced there. Eight-inch and 5·9-inch shells dropped into the south-eastern corner of the village with alarming frequency, and the company in support had to be brought back to cellars nearer Fosse 11 de Bethune. All through the night of the 12th/13th the shelling continued, mostly with 4·2's, which fell in all parts of the village, and there was very little sleep for anyone in Maroc that night.

Lieut. H. J. Dunn had a most wonderful escape. It sounds very like a fairy tale, but is nevertheless quite true. A shell penetrated the outside wall of the house and exploded on the ceiling of the cellar in which Dunn was sleeping. Absolutely the whole ceiling collapsed into the cellar, filling it to two feet deep with debris, with the single exception of a portion of about eight feet by four feet in the corner directly over Dunn's bed. Those who saw the scene next morning will never forget the sight of that bed, absolutely untouched in one corner of the cellar, which everywhere else was covered with ruins. Dunn seemed little the worse for the experience except for the absence of collar and tie, which, with sundry other articles of kit, were buried under the wreckage.

The next day there was considerable speculation as to the meaning of this sudden artillery activity, and the possibility of an enemy withdrawal on our front was considered quite likely in view of the salient which had been created between Vimy Ridge and Loos by the Battle of Arras. Throughout the day the enemy continued to shoot at our aeroplanes with rifles and machine guns, and did a certain amount of shelling. In the afternoon news was received that it was considered probable that the Boche had retired. Patrols from the front-line companies were immediately sent out to investigate, and these found the enemy trenches unoccupied except for small parties evidently left behind to shoot at aeroplanes and generally give the impression that the line was still held. The B Company patrol, under 2nd Lieut. G. S. L. Hayward, captured two prisoners of the 35th Fusilier Regiment, and the D Company patrol, under 2nd Lieut. D. F. W. Tanner, was engaged with and drove off an enemy rearguard. Orders were at once given for the front-line companies to move forward and occupy the front and support lines, which was accomplished by dusk. Capt. T. P. P. Walker and 2nd Lieut. K. Pfeuffar, both of C Company, leading their company over were wounded by a Boche sniper. When the enemy knew we had occupied his front-line trenches, he strafed pretty heavily for about an hour. Patrols were pushed forward and encountered no opposition, but orders had been received that no further advance was to take place for the present. Battalion Headquarters and A Company from Maroc moved up to our old front line in the course of the evening, while during the night the three front companies occupied the enemy's third line of trenches, known as Corsair, Cord, and Corporal. The night was very dark, which made moving about in unknown country difficult, particularly on the right and in the centre among the ruined buildings of Fosse 16 and the adjacent cottages. The occupation of this line represented a maximum advance of about 1,100 yards on the left and about 800 yards on the right.

Patrols continued active, and penetrated very much further into the enemy lines, but we had orders not to advance from the positions we had reached. The enemy kept up spasmodic shelling all night, but apparently did not know where we were, and a lot fell short of our lines altogether.

The next morning the 8th Queen's passed through us and continued the advance in Cité St. Pierre, meeting with only a little more resistance than we had had before. A Company moved into line with the other three companies, going into Corsair trench between B and D Companies, and Battalion Headquarters moved up into Fosse 16. The Battalion found working parties during the day consolidating a line of resistance which ran near the church in St. Pierre. Mindful of the experiences of others in the Boche retirement to the Hindenburg line earlier in the year, we were very chary about entering inviting-looking dug-outs, and treated new duck-boards and all movable objects and souvenirs with great suspicion. As it happened, however, the enemy had had very little time to prepare such surprises as booby-traps for us. Only one serious case occurred, in which a deep dug-out in the old enemy front line blew up and killed two men and wounded one. The Boche had made very extensive use of concrete throughout St. Pierre, and there were numbers of cellars so strongly protected that from our side they must have been nearly indestructible, but, of course, when we got them, the weaker sides and the entrances were facing the enemy.

Hostile shelling increased during the afternoon, and remained fairly active all night. In the early morning of the 15th, Battalion Headquarters moved up to a cellar just off Corsair trench. B and D Companies then went forward and sent patrols to try to make good the trenches between the two railway embankments east of the Cité. Cowden Trench and part of College Trench were occupied, but the enemy was found to be holding a strong point at the railway junction at the far end of College Trench. Late in the afternoon the 6th Division twice attempted to get forward on the left, and on each occasion the Boche gave us a very hot time, barraging Corkscrew Alley, the communication trench running along the railway, and strafing the village very severely. At dusk A and C Companies moved up to take over from B and D respectively.

Throughout the night the enemy shelled heavily, and as it was also very dark, the relief was not complete until daylight the next morning. Shells fell all over the village, but the enemy seemed to make a special target of the neighbourhood of the church, where A and B Company Headquarters and a support platoon of B Company were situated.

News was received at Company Headquarters about 5 a.m. that the relief of the forward platoons was complete. The dispositions were now A and C Companies in Cowden and College Trenches, C Company Headquarters being in a tunnel under the northern railway embankment, and A Company Headquarters in a tunnel under the southern embankment; B Company in support were in the trench near the church, which had been so badly shelled during the night; and D Company were in reserve near Battalion Headquarters.

It was soon after the completion of the relief that a very sad event occurred. The enemy were shelling with 5·9's the northern railway embankment and neighbouring trench, which was our most advanced position, and the tunnel which served C Company as headquarters was blown in. The tunnel appeared fairly safe, but although there were a good ten feet of embankment above it, it was only made of slag, and it turned out to be a regular death-trap. 2nd Lieut. R. C. Vaughan, commanding C Company, and 2nd Lieut. N. T. Johnson, of A Company, Coy.-Sergt.-Maj. Verrall, of C Company, and three men of the Company Headquarters were killed, and 2nd Lieuts. Lanaway, Holmes, and Woolley were dug out wounded. The last-named died of his wounds the next day. It had taken the whole day and the assistance of skilled miners to extricate him, and throughout that time he was perfectly cheerful and hopeful. He set a wonderful example of courage to everyone, and it was not until he was practically safe that the reaction set in and he seemed to collapse. We could ill afford to lose such officers as Vaughan, Johnson, and Woolley, and Coy-.Sergt.-Maj. Verrall was one of the best and most popular of warrant officers, and had been with the Battalion since its formation. The padre, the Rev. G. D. Barlow-Poole, came up in the afternoon and read the burial service, and a memorial tablet was subsequently erected.

All the officers of C Company were now casualties, and 2nd Lieut. Wade, of A Company, took temporary command of C.

At about 8.30 a.m. orders were received by the front companies to push forward patrols at 9 o'clock and try to make good more ground, and Maj. Whitty came up to establish advanced Battalion Headquarters with A Company and superintend the operations. The patrols found that the enemy was still holding strongly his position at the railway junction, and the houses to the north of the railway embankment contained snipers and machine guns. It was quite evident that the enemy had carried his voluntary retirement as far as he intended for the present, and that no further advance could be made without an organised attack, which was contrary to our instructions. A bombardment of the positions known to be held by the enemy was therefore arranged in the hope that he would evacuate them, but though patrols were able subsequently to penetrate to the north of the railway embankment and examine Puits 12 bis and the neighbouring buildings, the strong point remained held, and with the sniping continuing from the houses, it was found impossible to make good any more ground. The trench west of the railway strong point had been filled in, and Cpl. Kirby twice led patrols over this exposed part; but the enemy was watching it very carefully, and they were driven back by rifle and rifle-grenade fire. The shelling of St. Pierre continued spasmodically throughout the day. 2nd Lieut. E. B. A. Harvey, of B Company, was killed by a shell near the church, and Capt. Lock, Officer Commanding B Company, was wounded slightly for the second time in three days, but remained at duty.

The A and B Companies' Headquarters of the night before in the cellar near the church, which B Company had retained throughout the day, was blown in completely late in the afternoon, but fortunately it had been vacated only a few minutes before, as things were becoming altogether too hot there. Throughout the day fires and explosions were going on in Lens, which showed that if he did not definitely intend to retire further, the Boche was at least preparing for that possibility.

At 9 p.m. relief was commenced by the 9th East Surreys, and, owing to the somewhat complicated situation of the companies and the darkness of the night, combined with a considerable amount of shelling, it took all night, as before. The Battalion returned to support in the positions it occupied on the first night of the advance in Corsair and Corporal Trenches, and had fairly good and very welcome cellars and dug-outs. The next day was spent in consolidating the positions we were occupying as a second line of defence and carrying up ammunition and bombs to make forward dumps. Shelling was again fairly heavy, but not quite so bad as on the previous day. The 18th was also spent on consolidation, interrupted by considerable shelling, and in the evening a rather unexpected but very welcome relief turned up, the 4th Battalion Leicester Regiment of the 46th Division taking over from us. We were by no means sorry for the relief. We had been in the line, either at Maroc, Calonne, or in the advance, since March 26, that is twenty-three days, and if the last five days had been child's play to the Battle of the Somme, or to what we were to experience in the summer campaign, it had nevertheless been a rude awakening from the comparative quiet of the winter. The shelling was severe nearly the whole time. The five days cost us four officers and fifteen men killed or died of wounds, and five officers and forty-three men wounded; while the casualties for the whole of the six weeks at Maroc preceding the advance were only four other ranks killed and twenty-six wounded. We were strongly of the opinion that had we been allowed to push on further on the first and second days of the advance, that we could certainly have gained more ground and probably have taken Lens. As it was, after that time the enemy had strengthened his positions and was able to withstand anything short of an organised attack. The policy of caution, however, was doubtless due to the fact that the Division had all three brigades in the line occupying an extended frontage of about six miles, and with very little artillery supporting it on account of the demands of the Battle of Arras.

On relief we marched via Maroc and Bully-Grenay to rather crowded billets for the remainder of the night at Petit Sains (Fosse 10), which was reached about 3 a.m., and where the enemy greeted us by shelling the fosse.

PHASE V

THE SUMMER CAMPAIGN IN THE YPRES SALIENT

April—September 1917

AT 1.30 p.m. on the day after our relief in the Cité St. Pierre we marched off to Lozinghem via Hersin, Barlin, Haillicourt, and Marles-les-Mines, a distance of about eleven miles. Lozingham was quite a pleasant little village and the billets were good, but we only spent one night there, and the march was continued at 9 a.m. next morning to Ligny-les-Aires, which was also a good billeting area, and where we had five days in which to clean up.

At 9 a.m. on April 25 we marched to Delette, which was to be our training area. The billets were not too good for a rest area, but the ground available for training was suitable and extensive, and the country was charmingly fresh and open after the many months we had spent amongst the closely grouped mining villages and the ugly mine buildings of the Lens area.

In the morning the training was almost entirely devoted to open fighting with half an hour's close-order drill thrown in, while the afternoons were devoted to physical training, bayonet fighting, and training in Lewis gun, bomb, and rifle bomb.

Maj. C. P. Kingsland, from the 4th Territorial Reserve Battalion, who joined the Battalion when it came out of the Cité St. Pierre, was now commanding D Company, and Capt. H. L. Lewis, after his short stay with D Company, had returned to command C Company. Capt. H. S. Brown and Capt. W. G. Lock were commanding A and B Companies as before.

A good deal of football was played in the evenings. In the first round of the company competition, D Company beat A, and C Company beat B, both fairly easily. The final was played on May 5, and a most exciting game ended in a goalless draw. It was not possible to replay the match until the following September, when the two teams were very much altered by the ravages of the summer campaign, and C Company won by six goals to one.

On May 2 the Corps Commander, Lieut.-Gen. Sir Claude Jacobs, inspected the companies at training in the attack, and in the afternoon all officers of the Division went over to Corps Headquarters, where the Corps Commander shook hands with each one, and made a short speech, which gave us some inkling that we were in for a pretty warm summer.

On the same evening a battalion boxing competition was held. Excellent progress had been made by the competitors under the expert tuition of Pte. Bill Ladbury, of D Company, a former champion of England and holder of the Lonsdale belt. Ladbury himself gave a very clever sparring exhibition.

On May 7 the first blankets and winter clothing were withdrawn from the men, and officers had orders to reduce their kit to the regulation 35 lbs., and were also issued with privates' uniforms, so we had very little doubt left that there was going to be a war somewhere, and that we were going to be mixed up in it.

The last few days in this area were spent in practising trench-to-trench attacks, following a creeping barrage.

The weather was brilliantly fine during our rest, and on the whole it was very enjoyable. There can be no doubt that everyone left Delette feeling thoroughly fit and quite ready to face whatever the gods and the higher command, not to mention the Hun, had in store for us.

At 9 a.m. on May 10 we left Delette, and marched via Therouanne and Rebecq to Wittes, arriving there at 4.10 p.m. It was very hot, and a two and a half hours' halt for dinners was made in the middle of the day. Next day we remained at Wittes, and the Divisional Commander, Maj.-Gen. J. E. Capper, C.B., inspected the Battalion, and made a speech, complimenting us on what we had done in the past, and saying that even greater things would be required of us in the future, finishing up with blood and hatred in his very best style. This was the last time that Gen. Capper inspected the Battalion, as within a few days he left the Division, which he loved so much, to take charge of the Tank Corps, and we lost a fine soldier and a great friend.

The next day we had a long and exhausting

march in very hot weather via Steenbecque, Hazebrouck, and Steenvoorde to Beauvoorde in the Boeschepe area. We marched off at 5.30 a.m. and arrived in at about 7 p.m., after covering twenty-two miles. There was again a long midday halt for dinners.

At 10 a.m. on the 13th the march was continued via Poperinghe to Brandhoek. It was again very hot, and after arriving at 4.30 p.m., and getting into Dominion Camp, which had been taken over by the advanced party, we had to pack up again and move into Eerie Camp at 9.30 p.m. On that day, which was a Sunday, an advanced party of commanding officer, officers commanding companies, and others went by lorry from Beauvoorde to Ypres to reconnoitre the infantry barracks and ramparts which were to be taken over the next day. The party had a most appropriate welcome to the celebrated city; the barracks were heavily shelled with 5·9's from very soon after their arrival for nearly two hours. At the end of that time the structure and appearance of the building had undergone considerable modifications, and the advanced party viewed without enthusiasm the prospect of spending a week there.

The next evening at 9.45 the Battalion marched out of camp and entrained at Brandhoek, proceeding by train to the asylum just short of Ypres and then going on, the companies to infantry barracks and Battalion Headquarters to the ramparts. Here we were in brigade reserve, relieving the 8th Battalion of the Yorks and

Lancs Regiment. Maj. Whitty, with thirteen officers and 230 other ranks, remained behind in the transport lines, and moved on the 15th nto Toronto Camp.

The Battalion spent six fairly peaceful days in Ypres. The city was frequently shelled, but little of it fell in the barracks. These were built almost square, with an open courtyard in the middle. The outside wall on the eastern and the inside wall on the western side were very badly damaged, but the northern and southern wings were in fairly good repair, and were the most occupied.

On the evening of the 20th the Battalion marched out of the Lille Gate and proceeded to relieve the 8th Queen's in the right sub-section Hooge 1, the most easterly point in the Salient, three-quarters of a mile south of the village of Hooge. A, B, and C Companies were in the front line in that order from right to left, with D Company in support. The trenches lay in Sanctuary Wood, with right and centre companies on the southern slopes of Observatory Ridge, almost the only piece of high ground held by us in the Salient, and of further importance on account of the sap-heads of the extensive Canadian tunnelling system in Crab Crawl and Winnipeg Trenches. A Company and the right of B Company were holding Cross Trench, but further to the left it was too water-logged for occupation, and the rest of B Company and C Company held Vancouver Trench as the front line. C Company was in very low damp ground,

BATTALION HEADQUARTERS IN THE RAMPARTS AT YPRES.

YPRES.
The Square showing the Cloth Hall.

but was a long way from the enemy, and escaped almost all the strafing which occurred on the right. D Company were on the reverse side of the ridge, with Fort Street as their battle position. The relief was a quiet one, and was complete by midnight.

On the 21st 2nd Lieut. Orchardson of D Company was wounded in the shoulder by a shell splinter, and both that day and the next there was a good deal of " minnying " on the two right companies. Throughout the morning of the 23rd the enemy carried out a Minnenwerfer strafe mainly on the right of D Company, and did considerable damage both in casualties and to the trenches. The right half of Vancouver Trench was obliterated, and Cross Trench was blown in badly near the junction with Crab Crawl, cutting off communication between the two. The B Company stretcher-bearers showed great gallantry in getting back the wounded from isolated portions of Cross and Vancouver Trenches across the open under heavy fire.

In the evening the Brigadier came up and ordered the repair of the trenches as quickly as possible; but even before he had left the trenches at 7 p.m. an exceptionally heavy bombardment was opened, affecting the right centre and support companies. The majority of A Company were able to get shelter in the Crab Crawl tunnel and B Company, with D, again came in for the worst of it. Crab Crawl, Winnipeg Trench, Vigo Street and Hill Street were immediately blocked, but Fort Street and the left of

Winnipeg remained open, and D Company there filled the gap which was left in the front line. The B.G.C., who only escaped the bombardment by a few minutes, is reported to have said that there would not be a man left alive in the trenches. The high-explosive bombardment lasted for an hour, and was then followed by shrapnel, and all the companies stood to in expectation of an attack. Our artillery, however, was putting down a beautiful counter-barrage, which must have seriously disorganised any intended attack, and certainly in the gathering darkness and heavy smoke no assaulting troops were seen by our men. Next day some new stick bombs were found near the evacuated portion of Vancouver Trench, from which it would appear that a small portion of the enemy did succeed in trickling through that far.

The bombardment stopped at 8.55 p.m., but was renewed at 9.30, mostly on the support and communication trenches, until 11 p.m. One company of the support battalion, The Queens, was sent up in close support, and one company to the Zillebeke Dug-outs. Our casualties were seven killed and thirty-one wounded.

Our trenches were in a terrible state, and one hundred men of The Queens were sent up at once to assist us in digging them out.

The next evening D Company relieved B in the centre. Cross Trench was again occupied and communication with Crab Crawl restored. Vancouver was also occupied, except on the extreme right. To dig out this, a party of one hundred East Surreys were sent up, but they had considerable difficulty in retracing the line of the old trench. We were also hard at work all night, and a strong patrol from A Company, under 2nd Lieut. Griffith, was sent out to cover our working parties. The enemy, however, made no attempt to molest us, and had evidently suffered considerably himself, as he could be heard working all night.

On the night of the 26th/27th we were relieved by The Queens and returned, excepting A Company, who remained in close support in Wellington Crescent and Ritz Street, to the infantry barracks at Ypres without further casualties.

On the night of the 28th/29th we were relieved by the 2nd Bedfords, and proceeded by train from the siding near the asylum to Hopoutre, and from there we marched to camp near Wippenhoek. The area round the station and asylum was heavily shelled while we were entraining, and we had a most unpleasant finale to the tour. Lieut.-Col. Parker was slightly wounded, but remained at duty.

The Battalion arrived in camp in the early hours of the morning of May 29. The officers were quartered in some old and rather dilapidated huts and the men were under canvas which had been erected by the details at Transport Lines the day before.

That day was spent in cleaning up, kit inspection, and the posting to companies of a new draft of about seventy-five other ranks, which had joined the details a few days previously.

For the next few days training was carried out as far as was possible. Time was also spent in preparing the equipment for "Fighting Order," and collecting and practising with the various forms of ammunition, bombs, rifle grenades, flares, Lewis-gun drums, and other appurtenances of war which have to be hung on the person of a soldier about to enter into offensive operations.

All this time the guns were roaring away at their work of demolishing those celebrated defences, thought at one time to be almost impregnable, which protected the famous Messines-Wytschaete Ridge.

At about this time we lost two of our company commanders. Capt. Sir Henry Hawley, Bart., had served with the Battalion before it came overseas, and rejoined early in 1916. He succeeded Capt. T. P. P. Walker in command of C Company, and held this appointment with great success for some months. Capt. W. G. Lock joined D Company as a 2nd lieutenant, after the Battle of Loos, and succeeded Capt. C. F. Penton in command of the Company when the latter was wounded on the Somme. He was subsequently transferred to command B Company, and remained with them for eight months until he went down for a rest as a temporary instructor at a Musketry School; there he seems to have made himself indispensable, and he never came back. We were very sorry to lose them.

On the night 5th/6th the Battalion moved from the Wippenhoek Camp to N Camp, near Scottish lines at Busseboom, arriving in at 2.15

a.m. on the 6th. There were great numbers of troops and transport on the road and progress was very slow. The next day was spent in camp, and the Battalion again moved forward at 11 p.m., and went into billets in the ruined houses and cellars at Dickebusch, which was reached at 2 a.m. This day, June 7, was the day of the great 2nd Army offensive, known as the Battle of Messines. The 24th Division was in Corps Reserve to the 10th Corps, which was the left of the three corps engaged in the attack. The 72nd Brigade had no definite rôle allotted to it for the battle, but was ready to move off at a moment's notice if required to fulfil any of the duties which may be demanded of the reserve. The 17th and 73rd Brigades had objectives allotted to them, which they would be required to take in the event of the original attack proving a complete success. As it turned out, the attack was a success, and these two brigades moved up at about 3 o'clock in the afternoon, passed through the troops who had attacked in the morning, and successfully captured the final objectives with remarkably few casualties.

At about the time we reached Dickebusch the guns, which had been roaring so incessantly for many days past, seemed to grow strangely silent, and from half-past two onwards there was a hush which even the Boche must have felt boded him no good. This silence was broken as far as we were concerned just before 3 o'clock by the enemy himself, who began to shell the village with gas shells. Rumour had it that zero hour would be 3 o'clock, and when that hour came and nothing happened, many of those who had stayed out to see the mines go up and the opening of the barrage, went in to get a little rest. As a matter of fact, it was ten minutes later that great sheets of flame suddenly leaped from the earth, showing the positions of the nineteen great mines, the firing of which marked the opening of the second big offensive of 1917. Simultaneously thousands of guns opened along the whole of the nine miles' frontage of the attack. It was a most impressive sight, and it is small wonder that the enemy's front lines offered very little resistance to our attack. It is no exaggeration to say that not a blade of grass was showing anywhere over which that barrage had passed.

Our own part in this great battle was destined to be a passive one. So successful was the attack, following almost to the minute the exact time-table prepared for it, that the reserve was not called upon. Throughout the day prisoners and walking wounded came streaming down the track, which passed close to the village, the former looking very dejected and the latter equally elated. From these we learnt that everything was going well, and that our casualties were not very great, and those mainly slightly wounded.

At 1 p.m. we left our billets and marched up to trenches known as G.H.Q. 2nd line, just east of Café Belge, on the Voormezeele–Vlamertinghe road. Headquarters, A, B, and C Companies were in the neighbourhood of a large dug-out which was well supplied with sleeping accommodation, but which was so damp as to floor, walls, and ceiling, and smelt so foully, that the majority preferred sleeping under the open heavens close by. The disadvantage to this was a battery of particularly noisy 60-pounders firing directly over our heads from a distance of about 200 yards away. Those who have tried it will know that 60-pounders, even at a distance of half a mile, are not conducive to sleep. D Company found themselves some disused gun-pits on the other side of the road, and occupied them. Late in the evening and on the next day there was some shelling of the area round Café Belge, but though several shells fell in the actual field in which we were, they did not cause us any casualties.

At 9 p.m. on the 9th we marched up to the Brigade Reserve positions of the Mount Sorrel Section, relieving the 9th K.O.Y.L.I. and the 10th N.F.'s of the 23rd Division. A and B Companies were in Railway Dug-outs, and further forward Headquarters were in Rudkin House, and C and D Companies in Maple and Stafford Trenches, between those two noted spots, Maple Copse and Armagh Wood. The night was fairly quiet, but just after midday on the 10th the enemy shelled Stafford Trench heavily from the north-east, getting it in enfilade and giving the garrison a most unpleasant time. There were several casualties, including 2nd Lieut. R. N.

Killick, D.C.M., of C Company, who was badly wounded in the chest, and about six men of D Company, who were buried and were suffering badly with shell-shock when dug out. That night the Battalion relieved the 11th Sherwood Foresters in the left sub-section of the Mount Sorrel section. The journey up was a most unpleasant one. Either the Boche was attacking or thought that we were, as our front line and all communication trenches and approaches were plastered heavily for some time just as the Battalion was half-way up for the relief. The companies were absolutely caught, and for some time could neither move forward nor find any cover where they were. However, things quietened down, and the relief was eventually over by 2.30 a.m. This sector was the extreme left of the front of the attack on the 7th. D Company on the right in the "Triangle" and C Company in the centre were in the old Boche trenches, while A Company on the left was holding G Sap, which ran across what had been No Man's Land, and our own front line at Hedge Street. B Company were in support in Canada Street Tunnels. Headquarters were in Hedge Street Tunnel, a fairly extensive underground system, in which was also quartered a company of the 9th East Surreys in close support.

The trenches were, of course, in a very bad state indeed, particularly on the front over which the barrage had passed. The trenches there were almost obliterated, and, except in a few places, offered practically no cover.

At about the zero hour of our big attack on the morning of the 7th the enemy had made a determined raid on the left of our front, supported by heavy artillery and trench-mortar fire, which had caused considerable damage to these trenches. It also accounted for a very large number of dead Huns lying about the neighbourhood, and as they were now four days old and the weather was hot, they were not the most attractive feature of the locality. All these bodies had, of course, to be buried, an unpleasant job, and one which took a lot of labour which might very profitably have been spent on improving the trenches. G Sap, which connected C and A Companies, was only fully dug for half of its length on the A Company side, and this had to be continued, a task which was shared by A Company and the company of East Surreys in close support. Considerable improvement was made in all the company areas during our stay, but it would be idle to pretend that they were anything approaching good even when we left. Work was not easy on account of the frequency of the shelling on our whole front, which not only interfered with the work in progress, but also was continually causing further damage to the trenches.

One of the worst features of this sector was the difficulty of communication with the back areas. The communication trenches were very badly knocked about, and offered little cover either for safety or from view, and both they and overland ways of approach were frequently shelled, especially throughout the hours of darkness. This made the evacuation of wounded a very difficult matter; even when the party was not being shelled, its progress was very slow on account of the state of the ground. The ration parties also had very nasty journeys to and from Valley Cottages, where the rations were dumped. Valley Cottages itself was no paradise, and the journey also included such notoriously evil spots as Armagh Wood, Stafford Trench, and Rudkin House. The transport had no enviable task in bringing up the rations, the back areas coming in for a lot of attention by the Hun.

Most of the shelling on this front came from the north-east, and got our trenches absolutely in enfilade. During the day it was spasmodic; at times it would be quite quiet, and then a sudden burst of artillery fire would create a small hell for some minutes. But it was in the evening and early part of the night that the shelling rose to a crescendo. The enemy was extraordinarily jumpy; he seemed to expect us to continue our offensive, and was in the habit of putting up his S.O.S. at the least provocation, with the result that our guns, seeing his barrage and suspecting a counter-attack, would join in the disturbance, which would probably last about half an hour before both sides eased off gradually.

L.-Cpl. J. C. Bolton and Pte. J. H. Wheeler greatly distinguished themselves during this tour in maintaining communications under heavy fire, and both were awarded the Military Medal, and Coy.-Sergt.-Maj. F. Greenaway and L.-Cpl.

S. Browne of A Company also did very gallant work, and won the D.C.M. and the M.M. respectively.

On the night of the 13th/14th we were relieved by the 8th Queen's, and took over the support positions from the 9th East Surreys. There was some shelling during the relief, but on the whole the early part of the night was rather quieter than usual. The positions occupied were Headquarters in Hedge Street Tunnel, C Company in Canada Street Tunnel in close support to the 9th East Surreys, who were in the line on the right, and the other companies in the Fosse Wood and Stafford and Observatory Trench areas.

The four days spent in support were certainly no rest. The shelling of Rudkin House, Stafford Trench, and Fosse Wood was very severe. On the second day A Company evacuated Rudkin House as, although it appeared fairly deep and had considerable accommodation, it was really by no means safe, and the place had become a regular shell-trap. The next occupants, the North Staffords, lost several men buried by shell-fire in similar tunnels on the other side of the trench.

On the evening of the 17th we returned to the left subsection, Mount Sorrel Section, relieving the 8th Queen's, and were if anything rather glad of the change. The companies were disposed exactly as in the previous tour. Although we were only in the line two days this tour, it was quite sufficient. Soon after the completion of the relief at about 2 a.m. the enemy strafed very severely, mostly on the right company's front. D Company had a very bad time, and lost eight men, killed, including Pte. Bill Ladbury, the champion boxer, and three wounded. The rest of that day, the 18th, was very similar to the previous tour in the line, with intermittent shelling during the day, becoming heavier towards evening.

One of the features of both this and the former tour in the Hedge Street trenches was the activity of the enemy with low-flying aeroplanes over our lines. Our own aircraft had been very much in evidence actually at the time of the battle, and it was very seldom that a Hun plane was seen, but later he had things very much more his own way. He always appeared twice a day, at dawn and just before dusk, and flew so low that we used to say that he came in the morning to count the men in our trenches and again in the evening to see that no one had left.

The next day was fairly quiet during the morning, but in the afternoon the two left companies got badly shelled, and 2nd Lieut. G. H. Griffith, of A Company, and 2nd Lieut. S. A. French, of C Company, were wounded, in addition to other casualties. Things calmed down again in the evening, and we had hopes of a quiet relief, but our hopes were not to be fulfilled. Though in the forward areas it remained fairly quiet in the early part of the night, the ominous sound of crumping could be heard further in rear, and we realised that the relieving battalion, the 1st Royal Fusiliers, were having a nasty journey up. The relief was expected quite by 11 p.m., but midnight and 1 o'clock arrived without its appearance, and we began to wonder if we should get out at all that night, as it would be dawn at a very early hour, and a relief in daylight in this sector would have been most unwise. However, just as we were beginning to reckon up what supply of water and rations we had for an extra day in the line, the relief turned up, and we were able to get out just in time. Probably no one will forget that journey out of the line. There seemed to be shells everywhere. The mention of the names of the two main routes will probably bring back tender memories to many. The more northerly way was from Hedge Street down Davidson Street, past Sanctuary Dump and Stafford Trench, along the southern and western edges of Maple Copse, down Zillebeke Street to Hell-blast Corner, in Zillebeke along the Promenade on the southern bank of the Zillebeke Lake to Transport Farm and the railway dug-outs, and then down the side of the railway to Shrapnel Corner. The other route was from Davidson Street down Observatory Trench to Fosse Wood and Junction Trench, and along Fosse Way past Battersea Farm and Manor Farm to Transport Farm, and then as before. It did not matter which way you chose; the advocates of each were loud in their claims that their route was the worse. It was only too evident that some of the units going up for the relief had suffered very heavily. One very gallant company-sergeant-major, to whom apparently the particular sensation was new, was heard murmuring on the way out: "My God, what an awful thing it is to have the wind up."

Mic Mac Camp, for which we were bound, was a camp partly of huts and partly of canvas, situated between Ouderdom and Dickebusch, and some seven miles or more from the line, and when the Battalion dragged itself in there between 5 and 7 o'clock the next morning it was pretty well fagged out. The whole Brigade was here out of the line in Divisional Reserve. Four fairly restful days were spent in camp, during which the new Divisional Commander, Maj.-Gen. L. J. Bols, C.B., D.S.O., who was subsequently Chief of Staff to Sir E. Allenby in Palestine, inspected the Battalion.

The neighbourhood of the camp was sometimes shelled by a long-distance 9·5-inch gun. The shell made an unpleasant roar on arrival, likely to create alarm and despondency to those within some hundreds of yards, but never did us any real damage. Night bombing by enemy aircraft was incessant. The droning of enemy aeroplanes more or less directly over one's head throughout the hours of darkness on every night in which flying was possible—and at this time of the year that was nearly every night—was most unsettling, and robbed us of rest which was badly needed.

After four days at Mic Mac, on June 23, we moved up to Dickebusch, where we were in reserve to the rest of the Brigade, which had taken over the right subsection from the 73rd Brigade. The camp, which was of tents, situated to the east of the village near the brewery at Vyjverhoek, was known as the Brewery Camp. We spent three days in this camp, and except for the first night were not seriously molested, although the usual night bombing continued.

The camp was on the side of the Dickebusch–Café Belge road, along which there was a fairly deep dry ditch, and in this ditch the companies had been ordered to take cover in the event of shelling, and we had, during the first night, plenty of practice in moving from the camp to the ditch. Every hour from 11 p.m. until 3 a.m. the Hun dropped shells from a smallish high-velocity gun with considerable rapidity into the camp for about five minutes, just long enough in fact to get everyone thoroughly disturbed and all but the most bold into the ditch.

On the evening of the 26th we went up to relieve the 8th Queen's in the left subsection of the Battle Wood Section. It was a very wet day, but otherwise the relief was not particularly bad. The three companies in the line from the right were B, C, and A, with D Company in support. The line, which ran just on the eastern outskirts of Battle Wood, was held by means of short isolated lengths of trench garrisoned by either one platoon or two sections. Some of these trenches were freshly dug in the open, while others were parts of old Boche trenches repaired and fire-stepped for our own use. By day there was no communication between the forward posts, which were in the open clear of the wood, but as the enemy posts were fairly close to them, they escaped most of the shelling. The support platoons of the front line companies and the support company came in for far more shelling. The supports of A and C Companies on the left were in a ravine, along which the Boche had built some concrete pill-boxes. The two Company Headquarters shared one of these, named by the Boche Haus Barbara. The name conjured up visions of a charming little villa built in a wood, with beautiful green foliage and the birds singing in the trees, and perhaps a stream flowing lazily by. In fact, the villa was chiefly remarkable for a choice collection of fleas. The wood was in a very dilapidated state, and few of the trees stood more than ten feet high. The stream consisted of rainwater trickling from one shell-hole to another. "Ah," you ask, "but were the birds there, and did they sing ?" We are afraid that we cannot answer that question from memory, but we have it on the authority of one of our most eminent "Special Correspondents at the Front," that during one of the big battles the birds not only continued to sing, but chose the neighbourhood of the front line as a suitable place in which to do so; and in view of this incontestable (and doubtless first-hand) information, we are prepared to admit that it is possible that the birds did sing. Being truthful, however, we are compelled to admit that the only singing that impressed itself on our memory was that of the shells.

Haus Barbara had a direct hit with a 5·9 on the afternoon of the second day in, and stood the test well; a bit of concrete was chipped out of the face, and the occupants were mostly shaken on to the floor, while the inside was filled with dust and black smoke, but no one was hurt.

There were no communication trenches left in existence, and the tracks used, which in some places were in full view of the enemy, got badly strafed, while Battalion Headquarters in a pill-box in the wood also came in for a lot of attention, and Cpl. Littleproud, claimed to be the best officers' mess cook in the Division, was badly wounded in the leg. The Signallers also had a bad time, the telephone lines forward to the companies were continually being cut, and the linesmen had to spend a great deal of time on their dangerous job in most exposed positions. Among others, L.-Cpl. Bolton, an invaluable linesman with many months' fine service to his credit, was killed.

On the night 28th/29th the A Company post, held by 2nd Lieut. B. Wade and No. 2 Platoon, was raided by the enemy, who bombed it rather severely, but failed to enter the post, and were driven off by bombs and rifle fire with only one casualty to us. He, strange to say, was hit full pitch by a bomb which luckily did not explode. Lieut. Wade handled the situation with his unfailing resource and cheerfulness.

On June 27 part of the transport was sent off by road to the Lumbres rest area, while another portion followed by train on the 29th, and only such horses and vehicles as were necessary for the relief were kept back.

On the night 29th/30th we were relieved by the 8th Yorkshire Regiment, and we returned direct to Mic Mac Camp. The relief was complete at about 1 a.m., and although the prospect of our walk back had filled us with some thought, it turned out to be a comparatively peaceful night, and was probably the most quiet relief we ever had up in the Ypres area.

The casualties for the month were 23 other ranks killed and 3 officers and 53 other ranks wounded, and when one considers the number of shells and other forms of death which abounded, it is wonderful that the number was not much greater.

We got to Mic Mac Camp at 5 or 6 o'clock in the morning, and had a few hours to get a little rest and to get cleaned up. At 11.45 we marched in pouring rain to Reninghelst to entrain for Lumbres, whence we marched to Senningham, arriving in our billets at about 8 p.m.

Senningham was a pleasant little village, situated just south of the main St. Omer–Boulogne road and about three miles from Lumbres. The billets were not particularly good, but were considerably improved when we were able to procure some fresh straw for the floors. The first day or two were spent in cleaning up and kit inspections, and then training was started, chief attention being given to the prospect of future offensive operations. The ground in this neighbourhood was mostly covered with standing crops, and was not very suitable for training. Quite a good short rifle-range was made against the steep hill which sloped up to the main road, and rifles and Lewis guns could be heard cracking away there all day long.

Saturday, July 7, was a half holiday, and the Divisional Horse Show was held. The Battalion distinguished itself by taking the 1st Prize for travelling kitchens, C Company's cooker being the selected representative, while Diana, the C Company's charger, took 1st Prize for Infantry Officers' chargers.

The following Friday was observed as a whole holiday and Battalion Sports were held, organised by Maj. Whitty. One of the features was the officers' race of 100 yards, with handicap of 1 yard for each year of age over 21. Col. Parker and Lieut. Wood, the Quartermaster, who were on the forward marks, carried off the honours to the delight of everyone present. The tug-of-war was secured by C Company, and the most prominent in other events were Ptes. French, Lloyd, and Wells. The Transport supplied two most amusing events in the mule race and wrestling on mules.

The latter portion of the stay at Senningham was devoted to preparation for the next offensive, including the studying of a large model made by the Royal Engineers of the area in which the Division was to operate. The frontage of the

72nd Brigade was between the Mount Sorrel and Battle Wood Sectors, which we had just left.

At a very early hour on July 17, our stay at Senningham terminated, the Battalion marching away at 3.45 a.m. via Lumbres and Arques to Renescure, a distance of about fifteen miles. Packs were carried by motor lorry, and as the hottest part of the day was avoided by the early start, the march was not very exhausting. The billets at Renescure were excellent, and we could have done with a longer stay there.

The next day we left Renescure at 4.35 a.m., and marched via Staple and Hondeghem to Le Briarde, a shortish march of about eight miles. The next day we again had a short march to the area about Eecke and Steenvoorde, and an advanced party was sent on to prepare a camp near Reninghelst. At 5.5 a.m. on the 20th the march was continued, via Godewaersvelde and Boeschepe, to the Reninghelst area, where the camp consisted of a not too plentiful supply of bivouacs. On the 21st a further move was made at 7.30 a.m., via Ouderdom, to our old friend Mic Mac Camp.

On the evening of the next day we moved up and relieved the 10th West Riding Regiment of the 23rd Division in Brigade Reserve. Battalion Headquarters were in Larch Wood Tunnels, A Company in Battersea Farm, B Company in "the Dump," and C and D Companies in shelters at Scottish Wood.

On the 24th we relieved the 9th East Surreys on the 72nd Brigade battle front in Image

Crescent, with A and B Companies in the front line, C with Battalion Headquarters in Immovable Support, and D Company in reserve in Metropolitan Left. There was some gas shelling on the way up and during the next day. Mustard gas, which was at this time a fairly new form of frightfulness, was used plentifully in this area, but, thanks to good gas discipline, we were very free from casualties. One distinguished officer in the Brigade had the misfortune to sit down on ground contaminated with the stuff, with disastrous results.

The front line, which was, of course, an old Boche trench captured at the Battle of Messines, was still in very bad condition and was in places only three feet deep. There were a few old Boche concrete pill-boxes along it, but otherwise not much cover. It was in a small wood at the top of the eastern slope of the ridge, and afforded very good observation over the enemy country to the immediate front.

Our artillery preparation for the coming attack was steadily going on, and the enemy's guns were by no means silent, so that it was not very pleasant in or near the line. Knowing as he did that we were shortly going to attack, the enemy put down a protective barrage every morning from about half an hour before dawn until just after daylight, increasing it in intensity day by day as he grew more and more apprehensive. The quietest part of the day was from the cessation of the dawn hate until about 10 a.m., and then for the rest of the day the majority of the shelling would be searching for battery positions in rear, and the forward areas would be comparatively quiet until nightfall. Time was mostly spent in digging out and improving the assembly trenches, and in carrying up barrage rations and ammunition and bombs for forward dumps.

At 5.30 p.m. on the 26th a practice barrage was carried out by our artillery, and under cover of this 2nd Lieut. E. G. Brown, M.C., with eight men of B Company, carried out a raid on Job's Post, a strong point in a concrete pill-box between our line and the enemy's. The raid was a complete success, twelve men being captured, and 2nd Lieut. Brown was most deservedly awarded a bar to his Military Cross. An additional four Boches gave themselves up to B Company during the night. Cpl. G. Simmonds and Pte. R. T. Roper distinguished themselves in this raid, and won the Military Medal.

We had the misfortune on this day to lose 2nd Lieut. Wade, of A Company, with shell-shock. He was a very valuable officer under all conditions, and more especially so when things were most unpleasant, and the loss of his unfailing cheeriness was a great blow to his company.

On the night 27th/28th C Company relieved A Company in the front line, and the following night D Company relieved B. On the 28th we lost Maj. G. W. F. Powell, C Company, attached from 2nd/1st Kent Cyclist Battalion, who was killed by shell-fire after being with the Battalion only two months.

On July 28, during the enemy retaliation, a shell exploded at the entrance to a dug-out occupied by A Company, blowing a box of ignited hand grenades into the dug-out and wounding eleven of the twenty occupants. One bomb with the fuse burning fell amongst a group of men, and Pte. G. Neal, a young soldier (himself wounded), with the greatest courage and presence of mind picked up the bomb, carried it to the entrance of the dug-out, and threw it over the parapet, where it immediately exploded. Pte. Neal, whose gallant conduct undoubtedly saved the lives of many of his comrades, was awarded the Distinguished Conduct Medal, and was the first private soldier in the Battalion to earn that distinction.

At 4.20 a.m. on the 29th the enemy, seemingly looking upon it as a danger after his recent experience, blew up Job's Post. 2nd Lieut. Tanner took out a patrol in broad daylight, and in a very exposed position, to investigate what had happened, and showed great gallantry in getting back, under heavy fire, a man who was wounded in the open some way from our trench, and he was subsequently awarded a bar to his Military Cross. Sergt. E. Turner rendered great assistance, and was awarded the Military Medal.

That night the attacking battalions moved up to their assembly positions, C Company being relieved in the front line by the 1st North Staffords and D Company by the 8th Queens. The shelling was very severe all night, and relief was not complete until after daylight. All four companies were now concentrated in the Brigade support assembly positions north of the railway cutting, near Hill 60, in Immovable Trench and Immovable Support, the old enemy front and support trenches.

A Warning Order had been received that if the attack was a great success, the Battalion would be required to carry on the attack on the Brigade frontage, and capture the enemy positions to a depth of about 1,500 yards, and hold a line just east of Kent Farm.

On the 30th movement by day was restricted as much as possible, and work was concentrated on improving the existing dug-outs in our area. 2nd Lieuts. Proctor and Tanner, with the D Company carrying party, which was to accompany the attack, carried up the Stokes mortar ammunition to the assembly trench, where, as it happened, some of it was destroyed by enemy shell-fire during the night. In the early hours of the next morning, which was the day of the attack, the D and C Companies carrying parties moved up to their assembly positions, being all in at 3.10 a.m. The C Company party, under 2nd Lieut. L. P. Y. Smith, got badly strafed on the way up, but had no casualties. Heavy rain set in, and the going became very difficult and considerably interfered with the attack. Zero hour was at 3.50 a.m., and at that time the guide from the Stokes mortars, who was to take the carrying party, had not put in an appearance, and after waiting some time 2nd Lieut. Tanner went up, located the trench-mortar positions, and returned to take the carrying party up to the guns. 2nd Lieut. Smith was wounded, and there were casualties to about five other ranks.

The attacking battalions had very severe casualties, and a large party of D Company, under Coy.-Sergt.-Maj. Mitchell, who was awarded the Distinguished Conduct Medal, did fine work in helping to evacuate the wounded under heavy shell-fire. This work was carried on in most exposed ground on the forward slope of the ridge between Image Crescent and Jehovah Trench, the new front line. The wounded had been lying out there in pouring rain for many hours, and this rain, combined with the terribly ploughed up state of the ground, made the carrying of stretchers very difficult. Col. Parker afterwards received from the Commanding Officer of the Queens a letter containing his thanks for our great assistance in evacuating his wounded and for the unfailing regularity with which we had delivered his battalion's rations throughout the tour under circumstances which demanded the very greatest determination and devotion to duty. The doctor and his orderlies, assisted by Cpl. Andrews of A Company, also did very fine work.

On August 1 B Company moved out from Immovable Support into Metropolitan Trench to make room for the reserve company of the East Surreys, who relieved the Queens and the North Staffs in the front line. On the 3rd our Brigade frontage was extended to take over a large portion from the 73rd Brigade, and A

Company relieved the 13th Middlesex in Jeer Trench in Shrewsbury Forest. The front line here was fairly comfortable, and was well supplied with concrete dug-outs, but the journey up to or down from it through Shrewsbury Forest was absolute hell. Battalion Headquarters and D Company were in Canada Street Tunnels, which were in a state of appalling filth and horribly hot and overcrowded. C Company were in the old enemy front and support lines on Mount Sorrel, and B Company still in Metropolitan Trench. The weather continued bad until the 5th, then there was a slight improvement. On that day, after a heavy strafe, the enemy raided the East Surreys on our right, and came under heavy rifle and Lewis-gun fire from A Company. We were very sorry to hear that Col. de la Fontaine, the gallant Commanding Officer of the East Surreys, was killed in this affair.

During this tour Cpl. C. Andrews collected, dressed, and evacuated six wounded men of other units, who had been lying unattended in front of the line for three days.

The same night at 11 o'clock, just as D Company were starting out to relieve A, the enemy commenced a very heavy bombardment of the area, and caught D Company in the open. They had a very nasty time getting up to the front line, and had several casualties.

The 6th and 7th were fairly quiet days to what had gone before, and on the night 7th/8th we were relieved by the 17th Brigade, the 8th Buffs taking over the front line. The relief was none too pleasant, as the enemy shelled heavily, but eventually the whole Battalion was relieved by 4.30 a.m., and made its way out to Kruisstraathoek, where motor-buses were available for completing the journey to Mic Mac Camp. The tour had lasted seventeen days, and included a battle, with its attendant heavy shelling and exhausting demands for work and carrying, as well as some of the worst weather that has ever been experienced in mid-summer, and it is little wonder that the Battalion arrived in camp in a thoroughly exhausted condition.

We spent four days at Mic Mac cleaning up and resting. These were, on the whole, quite uneventful, except for enemy aerial activity at night. The continual droning of enemy aircraft all night and every night, mixed with three short blasts on the whistle at frequent intervals by the sentry and shouts of " Put that —— light out," and the sound of bursting bombs at varying distances, was our unkind fate. Drafts of 170 other ranks, which had arrived during the tour, were posted to companies.

On the 10th Maj. Whitty went into hospital with trench fever, and a few days later Capt. H. J. Wenyon, who had been attached to the staff since February, returned from Division to take over the duties of 2nd-in-Command.

The Division had now only one brigade in the line, one brigade in support at Dickebusch, and the third in reserve at Mic Mac Camp, tours being of four days' duration. On the 11th we moved up to support in H Camp at Dickebusch, where the accommodation in tents was rather crowded. Life there was very similar to that at Mic Mac, not a great deal of work was done, and night bombing was the rule rather than the exception. An additional diversion was offered when the enemy carried out shoots on the captive balloons in the neighbourhood, during which our camp was in quite as much danger as the balloons.

Just about 200 yards behind the camp there was a 9·2 naval gun on a railway mounting, which used to shoot at the enemy railhead at Roulers and at his billets in the back areas. Its rate of fire was usually only four or five rounds an hour, so that quite a small shoot kept the neighbourhood awake for an unpleasantly long time.

On the afternoon of the 15th the enemy carried out a most alarming daylight air-raid with six bombing machines on the siding between the camp and Café Belge. There were extensive horse lines all round, and the casualties to men and horses were severe. The place was a horrible sight just afterwards.

That evening we went back into the line, taking over from the 7th Northants the right subsection, Shrewsbury Forest, which was the original 72nd Brigade battle front. B Company went into the front line in Jehovah Trench, or rather in isolated posts in shell holes about fifty yards forward of where the trench had been, as in many places it was barely distinguishable. The weather was still very bad, and the whole area, which was absolutely churned up by shell-fire, was no better than a collection of little lakes, and shell-hole life was by no means a pleasure. However, the enemy was some way away, and did not trouble us with trench mortars and only very little with machine guns, and all the shelling was directed further back. C Company were in support in Image Crescent and Image Reserve. Battalion Headquarters was in a pill-box in the wood just north of the Zwarteleen–Klein Zillebeke road, and A and D Companies were in Immovable Support and Immovable Trench respectively. All these areas came in for a fair amount of shelling at different times, but the worst places were really further back in the battery positions, and such places as Larch Wood and Jackson's Dump; really in the line things were very much quieter this tour than they had been since our arrival up north in May.

On the 16th there was an attack by the 5th Army just north of us. The artillery on our front was all engaged in the barrage, but we did not come in for any of the counter-strafe.

On the next night C Company relieved B, two days being quite sufficient in the front line posts in this appalling weather.

On the evening of the 19th we were relieved by the 3rd Rifle Brigade and returned to Mic Mac. The early portion of the journey out was fairly quiet, but there was some nasty shelling by 8-inch howitzers in the neighbourhood of Lock 8 and for some little way further back.

The four days spent at Mic Mac were very similar to our previous visits, and on the 23rd we again went up to Camp H at Dickebusch. A large draft of officers arrived at about this time, bringing the numbers up to seven a company.

There was in close proximity to our own somewhat crowded camp of tents a most beautiful little camp made of very attractive huts, absolutely new and clean. This camp was labelled "Reserved for Corps Cavalry only." Now, it seemed a pity that the Corps Cavalry, who had not for months seen a vestige of a battle, and had been living in places of much greater comfort than the P.B.I., should have such a beautiful place reserved for them, when they were not even there to use it, and two Company Headquarters consequently decided to appropriate one of these huts for a joint mess. Shortly after dinner a senior officer entered and asked, somewhat peremptorily, "What are these lights doing here?" The answer came back pat from a most distinguished subaltern in the far corner of the hut, in the one word, "Burning."

At 5 p.m. on the next day we were suddenly moved from our camp to Château Segard in support of the 14th Division, who were reported to be in difficulties in Inverness Copse. It was said to be quite likely that we should be sent up to counter-attack, as the situation was "obscure." There was absolutely no accommodation available at all, and we had to make such bivouacs as we could with waterproof sheets and any R.E. material which was lying about. The next morning we returned to H Camp, as the situation had been cleared up without our assistance. That afternoon there was another big air-raid on the siding near the Brewery, and we know of at least one officer having a bath there at the time who very nearly left the building hurriedly in most unconventional attire.

On the 27th we returned to the old trenches, A Company being in the front line for the first two days, and D Company for the second. There was some heavy shelling on the way up in response to an S.O.S. on our left, but we were lucky in escaping casualties. The tour was very similar to the previous one, except that on the second day the weather at last really cleared after being at its worst for exactly a month, and that on the night 29th/30th 2nd Lieut. Carville, of D Company, collected a couple of Boches out of a shell-hole in front of our line of posts.

From Jehovah Trench it was by no means a pleasant journey back to Battalion Headquarters, as the way led up a steep slope for about 660 yards to Image Crescent, without a communication trench, in full view of the enemy, and with the ground so pitted with water-logged shell-holes that progress was bound to be very slow. One afternoon, after a perfectly quiet and uneventful day, in which there was absolutely nothing to put in the intelligence report, which was then due, the Company Commander in the front line decided, in order to save a runner the journey back with the report, to send it over the telephone by B.A.B. Code. This consisted of groups of figures representing sentences in general use. For instance, the sentence, "Our rum has not yet arrived," might be represented by the figures 528, and " Please expedite " by 172. The only difficulty was that there was sometimes a difference of opinion as to how to make the correction. In this particular case the message selected by the Company Commander was " Situation normal," which was duly dispatched to Battalion Headquarters. A few minutes later the Company Commander was rung up on the telephone by the Battalion Intelligence Officer. He gathered that his message had caused considerable alarm and despondency at Battalion Headquarters. He was able to reassure them until the report could be sent down by hand, and he then discovered that the message as deciphered at Battalion Headquarters read : " Tanks are approaching," which had not unnaturally caused a considerable stir, and the Commanding Officer was already on his way up to Image Crescent, armed with field glasses, to investigate.

On the 31st we were relieved by the 8th Buffs, and returned to Mic Mac.

On the 30th Maj. Whitty returned from hospital, and a couple of days later Col. Parker went down, after being in very indifferent health for some time. Maj. Whitty took over command of the Battalion, and Capt. Wenyon remained as 2nd-in-Command.

Four days were spent at Mic Mac, and four at Dickebusch as before. The weather was very much better, and night bombing very active. On the 5th Lieut. G. Bullen, who was commanding B Company, left to join the R.F.C., and Lieut.

E. G. Brown, M.C., took over command of the Company.

While at Dickebusch company commanders and others went up to reconnoitre what was to be our new front for the next tour, to wit, Inverness Copse. We had heard quite enough about Inverness Copse for some time past to make the prospect by no means alluring, in fact that celebrated spot, together with Glencorse Wood, figured almost daily in the communiqués as either attacked or counter-attacked, and more often than not the situation there was described as "obscure."

The way up passed a succession of places whose names will never be forgotten. First Ypres and the Main Square, then out through the famous Menin Gate and along the Menin road, past the Ecole on the right and the White Château on the left, the way led to Hellfire Corner and Birr cross-roads, Outpost Farm, the Culvert, and finally through what had been Hooge village. A journey up that long stretch of the Menin road was not undertaken without considerable fear and trembling, but we know of one company commander who went to reconnoitre the line held by the 2nd Leinsters, and who, after visiting each of the four company headquarters and battalion headquarters, and receiving typical Irish hospitality in the shape of a four by two (i.e. four inches of whisky and two inches of water) at each, came down the Menin road quite pleased with himself.

The relief took place on the morning of the

7th in broad daylight, and we have probably got the fact that it was rather misty to thank for the forbearance of the enemy in letting the whole Battalion march up the Menin road by platoons at intervals of 200 yards, almost without a shell. The Menin road was so heavily and persistently shelled by night, that to relieve by daylight and risk being seen was considered preferable to using it at night.

B, C, and D Companies took over the front line in that order from the right, B and C being south of the Menin road and D north of it. South of the road the main line of defence was New Cut —an incomplete trench which ran across Jasper Avenue and Jasper Lane to the Menin road, where work was in hand to connect it with the trenches north of the road. Forward of this there was a line of posts just inside the near edge of the copse, along the line of the old trench which connected Jasper Lane and Jasper Avenue. In the centre, C Company's posts were in communication along this trench, but the B Company's posts were isolated, the trench in the copse and that end of Jasper Avenue being both impassable. On the left D Company had a fairly normal trench system.

Battalion Headquarters were in a dug-out near Clapham Junction, and A Company in support were quartered in the long tunnel which ran under the Menin road. For the first two days things were fairly quiet, except for periodical shelling of the Stirling Castle neighbourhood, which was sufficiently near to B Company's

Headquarters and supports to be uncomfortable, and also of the entrances to the Menin Tunnel, which was apt to prove disconcerting to A Company.

At 4 a.m., about dawn, on the 9th, the enemy opened a very heavy barrage on our lines to a depth of several hundred yards, including Battalion Headquarters and the entrances to the Menin Road Tunnel. Under cover of this they attacked our advanced posts in the Copse with great determination, using specially picked and trained "Sturmtruppen" with Flammenwerfer. The barrage behind this line of posts and on the support line totally isolated them, and even at the tunnel of the support company the barrage was so severe that it was impossible to get a man out with a chance of reaching his destination.

On the right the attack was completely repulsed with Lewis-gun and rifle fire, assisted by our artillery barrage, with very heavy casualties to the enemy. This was largely due on the B Company front to the magnificent example and coolness of 2nd Lieut. A. B. Richardson, who was there in command, and the excellent behaviour of the men under him, and the right post of C Company, under Lieut. H. R. James, also put up a splendid performance. The repulse of the attack after a bombardment which might easily have proved demoralising to any but the stoutest troops, and under Flammenwerfer with which we had had no previous experience, and to face which is known to require a very high degree of courage, was a performance of which the B and C Company troops concerned had every reason to be proud.

The only post entered was that on the left of C Company, which consisted of a large concrete pill-box and the trench in its vicinity held by 2nd Lieut. H. W. Beattie and his platoon. The enemy were able to approach very close to this post under cover of the undergrowth and fallen trees, and although the platoon put up a stout resistance, the enemy got in and succeeded in capturing fifteen men, many of whom had been wounded in the preliminary bombardment. 2nd Lieut. Beattie succeeded in escaping, shooting two of the enemy with his revolver.

Immediately the news of the loss of the post reached Battalion Headquarters, Maj. Whitty gave instructions to Capt. Wenyon to go up and restore the situation. No. 2 Platoon of A Company was sent up to assist, and bombs and rifle grenades were rushed forward. Capt. Wenyon very quickly grasped the situation, and his preparations for counter-attack were rapidly made. The plan was of the simplest, and was consequently brilliantly successful. The attack was carried out by C Company, with two parties under 2nd Lieuts. Pfeuffar and Beattie, working from different directions, the covering fire being supplied by bombs, rifle grenades, and Lewis-gun fire at very close range. L.-Cpl. Cook, with the Lewis-gun section of No. 2 Platoon, worked his way to a very exposed position, within twenty-five yards of the pill-box, at considerable risk,

THE MENIN ROAD AT HOOGE.

INVERNESS COPSE.
Showing the concrete Pill Box recaptured on September 9, and the Menin Road

and was able to pour in a murderous fire which greatly assisted the attack. The fire of this Lewis gun, added to the very accurate rifle-grenade barrage, kept the heads of the enemy very much down, and enabled 2nd Lieut. Pfeuffar to lead his party right across the front of the dug-out and to launch his attack from the unexpected flank.

The attack, with 2nd Lieut. Pfeuffar and Sergt. Purfield very much to the fore throughout, was a great success, and resulted in the complete re-establishment of the line and the capture of fifteen prisoners, one machine gun, and one light trench mortar, while in addition the enemy casualties both in the attack and counter-attack were known to be heavy, and several of their dead were left in and around the post. The Lewis guns of D Company on the north side of the road did great execution and were very valuable. After the attack L.-Cpl. Cook worked his way with his Lewis-gun team along the embanked Menin road, 100 yards in front of our positions, and covered the work of consolidation.

The success of the counter-attack clearly demonstrated the value of immediate action being taken by the troops on the spot, based on a bold but simple plan, and supported solely by the infantry weapons, thereby saving the time required in making elaborate arrangements for artillery support, which the enemy would gain for consolidating and organising his defences. Capt. Wenyon was to be heartily congratulated on the result.

The following decorations were awarded for

this action : Capt. H. J. Wenyon, acting 2nd-in-Command, D.S.O. ; 2nd Lieuts. K. Pfeuffar and H. W. Beattie, C Company, M.C. ; and Sergt. Purfield, C Company, and L.-Cpl. Cook, A Company, the D.C.M.

2nd Lieut. H. C. Walton, of D Company, was killed in the bombardment, and Lieut. W. R. C. Hodge, also of D Company, was severely wounded. The casualties to other ranks during the day included twelve men killed.

While we were there the enemy made no further attempt to clear us from our footing in Inverness Copse, and the rest of that day and the next were as quiet as could be expected in such a very volcanic neighbourhood. On September 11 we were relieved in daylight by the 1st Battalion Royal Fusiliers, and we returned to Mic Mac without undue molestation by the Hun.

It turned out that we were about to shake the mud of Flanders from our feet for the last time, and it was with mixed feelings that we learnt of our impending departure for pastures new. It may surprise many of those who read the foregoing account of our stay up north to hear that there could be anything but joy at the prospect of departure. They read of shells and mud, of battles, gas, and air-raids, with particular emphasis on shells and mud, and it must be admitted that the dominant sensation on getting our orders for the move was one of joy—and yet somehow the life up there was not all bad. In spite of night bombing and long-range shelling there was something quite restful about Mic

Mac Camp and those camps at Dickebusch, and the good fellowship throughout the Battalion seemed to increase in the mud of those Ypres trenches. If the tours in the forward area were bad, the arrangement of reliefs was such that, at least for the two final months, we got more time out of the line than ever before or since, and so we were leaving evils which we knew for evils which we knew not, and there was at least a possibility that the last state might be worse than the first.

In the early morning of September 14 we got into motor-buses and lorries which took us to Merris, a very nice farming village six miles south-west of Bailleul. The billets, which were scattered about the various farms, covered an extended area. It was good to be out again in fresh country unspoiled by shell-holes, and the place had an air of rest and peace which was very pleasant. It is sad to think now of the state to which this area was reduced by the fighting of 1918.

First we set to work cleaning up, and then training was started, but not too strenuously. Some football was played, and a very good match between the officers of the Battalion and Brigade Headquarters ended in a draw, one of the features of the match being the goalkeeping of Gen. Sweny. A great rugby match was also played, the only one we remember in France in our Brigade, in which The Queen's Own and Brigade Headquarters beat The Queens, East Surreys, and North Staffords by one goal and one try (eight points) to nil.

On the 18th Maj. Whitty went on leave, and Capt. Wenyon took over temporary command. The Battalion was inspected on that day by Maj.-Gen. A. C. Daly, the new Divisional Commander, who succeeded Maj.-Gen. L. J. Bols, C.B., D.S.O., when the latter left to join Gen. Allenby in Palestine as Chief of the General Staff. Gen. Daly remained in command of the Division until it was disbanded in 1919. Capt. T. P. P. Walker here rejoined the Battalion, and resumed the command of C Company.

We finally left the area on September 21, when we marched out of Merris at 5.30 a.m. and entrained at Caestre, which we left about midday. B Company had gone to Caestre the day previous, and had been engaged in entraining the transport for the whole Brigade.

And so we left the Salient for the last time. There were few square inches of mud there from the Menin road southwards with which the Battalion had not been closely associated, either in the autumn and winter of 1915–16 or in the summer of 1917, and our experience had always been of fighting and shelling and mud, never of the more peaceful periods which were known in its history; and as our train moved out of the station at Caestre, we thought of the many good comrades who remained behind sleeping beneath the wooden crosses in Flanders' fields, which marked the places where the Battalion so dear to us all had fought and won; and we were proud that it was ours to hold high the torch which their hands had thrown to us.

PHASE VI

TRENCH WARFARE IN THE DEVASTATED AREA AT HARGICOURT

September 1917—*March* 1918

For a troop train journey, the one from Caestre was quite good, and we arrived at the desolate and demolished station of Miraumont in Picardy about 11 p.m., and marched to camps at Beaulencourt, just outside Bapaume. By no means all the roads in this neighbourhood were as yet repaired, and we had to make a long detour, marching in all eleven miles when the actual distance was only seven. We stayed here a few days, and many of the Battalion took the opportunity of visiting the old scenes at Guillemont and Delville Wood. We heard indefinite reports of the front at Hargicourt, which we were soon to take over. There had just been a battle there to straighten our line by taking the Cologne Ridge. The attack had been successful, and the front was said to be still noisy and unpleasant.

On September 25 we moved to camp at Haut Allaines, near Peronne. From here, on the 26th, parties went up to reconnoitre the positions at Hargicourt. On the 27th we moved by motor-bus to Roisel for the night, and on the night of 28th/29th the Battalion relieved the 9th Northumberland Fusiliers (34th Division) as support battalion.

Our Battalion Headquarters were at Cote Wood, and the Battalion was disposed in trenches running up hill and down dale in the region of Cote Wood and Fervaque Farm. The weather was beautiful, all was quiet, and we had an opportunity of reconnoitring the front line thoroughly. The whole Division was in the line here, the 17th Brigade on the right, 72nd Brigade in the centre, and the 73rd Brigade on the left.

We were to share the left subsection of the centre section with the 9th East Surrey Regiment. The 8th Queen's Regiment and the 1st North Staffordshire Regiment worked the right subsection. Our front section consisted of a front line—tumbling to pieces—called Railway Trench on the right, and Pond Trench and Bait Trench on the left, a support line just being dug about 200 yards behind, and reserve positions with Battalion Headquarters in a chalk quarry, known as The Egg, just outside the ruined villages of Hargicourt and Villeret. There were good communication trenches, Club Trench and Onion Lane.

From the front line at Bait Trench we had a splendid view of the Hindenburg line and the village of Bellicourt. The ground sloped down steeply from this part of our positions, and we could see almost the whole of the valley called Buckshot Ravine, across which the Hindenburg front line and switches in front of Bellicourt had been dug. The country behind our lines was undulating and splendidly open. All villages had been totally destroyed by the enemy on his retirement in the spring.

The position of the support battalion at Cote Wood was quite comfortable. Protected from view, the trenches were well dug on the reverse slope of a ridge.

Our transport lines and Quartermaster Stores were established some four miles back in an orchard and paddock close to an old destroyed sugar factory, known as Montigny Farm. There was a rest camp here for the battalion out of the line in Divisional Reserve. From time to time our quarters when in Divisional Reserve were changed. We started at Montigny, later we used the ruined village of Hancourt, and last we were very comfortably quartered at Vendelles. Great improvements were made in the accommodation by the 24th Division during its long stay in this region.

The weather was fine when we arrived at Cote Wood, and we immediately took advantage of this to set our house in order for the winter. We were up working at full strength on the communication trenches to start with. By strenuous efforts, before the weather broke, we succeeded in getting our front and support lines and communication trenches into a state which would hold up in the winter. In the support line we started making deep dug-outs on our own, and Capt. Proctor and 2nd Lieut. Holliday did great work in this direction. Everywhere was a scene of strenuous activity. Fortunately the Boche interfered very little with us during this period, and later on, when things livened up considerably, we reaped the benefit in the shape of dry trenches and comparative safety. Half-way through the winter we side-slipped to the left a bit, taking over a piece of trench from some other people who had not seen fit to take advantage of the dry spell in October. This piece of trench—Fish Lane locality—caused us more trouble than the whole of the rest of our line together. Down at Montigny Farm there was a lot of work for the transport and Quartermaster's Stores personnel. When they got there they found just a few huts and no covered standings for horses at all. Material for building was very difficult to obtain by indenting on the R.E.'s, and there was nothing for it but to " find " the stuff for building if we were going to make men and horses comfortable for the winter. Capt. Arnaud moved in a mysterious way his wonders to perform at Montigny. G.S. wagons and limbers used to creep out of the camp during the evening. No one knew where they were going or when they returned. But a growing dump of corrugated iron, timber, and bricks and some feverish activity in the building world might have given some clue to any one good at putting two and two together. By the importunity of our asking we finally got half an Adrian hut supplied, and made out of it a really splendid Quartermaster's Stores.

The N.C.O.s and men built their own shacks. Some of them were comical enough to look at, but they were pretty sound inside. Capt. Arnaud's transport lines were a triumph. He

MONTIGNY FARM.
The Sugar Refinery.

VENDELLES.
Occupied by the Reserve Battalion.

made really splendid covered standings for all the animals, brick walls, wooden stalls, brick floor, corrugated iron roof—all without an inch of material supplied by the R.E.'s. Montigny was visited by some of the Battalion in April 1919, and most of these splendid covered standings were still there.

In the line we had six-day tours and daylight reliefs. There was no very great excitement while we were getting to know the line. The enemy was, on the whole, very quiet, and seemed quite reconciled to the loss of his bit of high ground at Cologne Farm. On October 17 we side-slipped northwards, and took over the extra bit of line previously referred to. At the same time we had to give up our Battalion Headquarters at The Egg, where we had just constructed a new mess, a deep dug-out, and many shelters, and where we were getting comfortably settled. Was it not ever thus? We only moved about 500 yards to the railway embankment, where there were a few shelters known as the Leicester Lounge. By the railway there was the remains of an old engine, popularly known as the "Leave Train." We started building hard again here, and made good quarters, which we were fortunate enough to retain for the remainder of the winter. On the 22nd we moved back to Vendelles for six days, our first tour in Divisional Reserve.

The Battalion was housed in big Adrian huts, in rather a dilapidated condition. The village, of course, was totally destroyed.

There was very little in the way of amusement for anyone in this devastated country. Cinema shows, canteens, and concert huts were an urgent necessity, and the building of these was being taken up by the Division most energetically.

November 1917 was a month of changing fortunes, alarms, rumours, orders, and counter-orders.

On November 3 Lieut.-Col. J. C. Parker proceeded to England for a six months' tour of duty, and, as events turned out, this concluded his appointment, held for over two years, as our Battalion Commander. Col. Parker joined us from the 6th Battalion at the time of chaos just after Loos. His work in the muddy trenches at The Bluff, Canal Bank, and St. Eloi in October and November 1915 will never be forgotten. During the months at Ypres in 1916 the Colonel was really very ill, but he stuck it out then and right through the summer, until he was carried away at Fricourt with a very severe attack of malaria and trench fever. During the Ypres fighting of 1917 he was again very ill and very weak, but he refused to admit that anything was wrong. In the autumn of 1917 he had to go down for an operation, and when he returned in November he still looked so ill that the Brigade Commander insisted on his having a rest. And so in a quiet manner he left the Battalion which he had worked for with infinite bravery, and which he had loved so well for two long and arduous years. Col. Parker had led the Battalion successfully through very many

THE COTE WOOD POSITIONS.
Occupied by the Battalion in support.

"LEICESTER LOUNGE."
Front line Battalion Headquarters, showing the "Leave Train."

of its most difficult experiences. All those who were intimately associated with him in his work knew and will always remember the great things he did for the 8th Battalion Royal West Kent Regiment.

In the absence of Maj. Whitty, Maj. Wenyon took over command temporarily, when Col. Parker left us.

During the winter we concentrated on work on and in front of the front line. Organised wiring on an extensive scale was taken up, and we started digging a new trench in front of Pond Trench in order to cut off the awkward and waterlogged Fish Lane positions recently taken over. On November 20 the 9th East Surreys did a raid in the early morning. Demonstrations and raids were taking place all along the front that morning in conjunction with the great surprise attack on Cambrai. In the afternoon we relieved the East Surreys in the front line, and every night we kept the closest possible touch with the enemy by continuous patrolling on a large scale, in order to find out if he was showing any signs of retiring opposite us as a result of Cambrai.

In the early hours of November 26, Lieut. H. J. Dunn, the Battalion Intelligence Officer, was shot dead when he was out with a party digging our new advanced trench. It was really too bright a moonlight night to go out in such an exposed position, but, knowing how urgent the work was, Dunn went out with two men to see if he could possibly get the party out to their work in safety. He was shot through the head almost immediately. Dunn was a splendid fellow. He took the most devoted interest in his work, first as Platoon Commander at Hulluch, and later until his death as Battalion Intelligence Officer. In this capacity he was to be seen at all hours of the day and night wandering round in the front line with compasses, field glasses, map cases, sniping rifles, and every modern device for accurately surveying and killing unwary Huns. He was a very talented musician, and withal he was such a cheery, lovable man. When he went, everyone in the Battalion felt he had lost a great personal friend. He was buried at Hargicourt cemetery in the afternoon. Major Whitty rejoined the Battalion, and took over command that day, and his first duty was to attend Dunn's funeral.

The position of Battalion Intelligence Officer was taken over by Lieut. G. S. Bowen, of A Company. To take over and carry on Dunn's work was no easy job, but Bowen tackled the job well, and he was still I.O. when the Armistice was signed.

The Division was to have been relieved at the end of November, but on the 30th the counter-attack at Cambrai took place, the relief was cancelled, and it looked as though we might be in action with the Hun at any minute. One battalion of our Brigade was hurriedly sent out to Ronssoy to protect our flank in the direction of Epèhy, and for a few hours the situation was decidedly obscure. Things blew over later, and on December 1 we relieved the East Surreys again in the front line, resuming normal trench life. Things were beginning to wake up a bit, however, and the discussion of detailed defence schemes—which was later in other sectors to become the mania of the higher command—started to take prominence.

During December the newly dug support line was strongly wired. Gas projector attacks by the Special Company R.E. took place on a biggish scale. On December 8 2nd Lieut. Carville and 2nd Lieut. Holliday made a great effort to get through the enemy wire with a small party. They blew up the wire to a depth of fifteen feet with a Bangalore torpedo, but there was another fifteen feet to blow. Holliday was nearly blown up himself in setting off the torpedo. They struggled hard to get through, but the Hun had been very good with his wiring here, and after getting their clothes very badly torn, the party had to come back. After a spell at Vendelles again, when most of the Battalion were out daily on road making at Jeancourt, we were back in the line again on the 14th. 2nd Lieut. Holliday was wounded on the 16th when out with a wiring party.

A spell of very cold weather set in, thick snow alternated with hard frost for nearly a month. About this time the Division held the line with two brigades, and one brigade was out.

On December 21 we were relieved by the 8th Buffs, 17th Brigade, and moved out to Hancourt. Here Christmas Day was spent. The accommodation at Hancourt was not the best in the world, but each company had its Christmas dinner, and the whole Battalion had a happy but quiet Christmas. On the 26th Lieut.-Col. N. I. Whitty, D.S.O., was suddenly ordered to proceed to England to take over command of one of the newly formed Divisional Machine-gun Battalions. When he returned to France a few weeks later he was posted to our Division, and we often saw him, and frequently did stunts together in which he assisted us with his guns.

He had been with us for two years. He was 2nd-in-Command under Col. Parker, and he commanded the Battalion for about seven months when Col. Parker was away. His energy while he was in command of the Battalion was tremendous. At all times of the day and night he was up inspecting the work in the trenches, planning and building dug-outs, organising artillery shoots and raids, salving material, opening canteens in the trenches, and a hundred-and-one other things. He was a great worker for the 8th Battalion. His direct forceful way of speaking and acting appealed very strongly to everyone. The Battalion owes a very great deal to the work of Col. Whitty.

He was awarded the D.S.O. in 1917, in recognition of his splendid work on the Somme, and in the command of the Battalion at Hulluch during the winter of 1916–17.

About this time our medical officer, Captain G. A. McLarty, left us. He is one who will always be remembered in the life of our Battalion.

Joining us at Wulverghem in March 1916, almost his first experience was the bombardment of Tea Farm. On this occasion he worked magnificently amongst the shells until the last wounded man was got away. Under fire he was always splendid. His great work on the Somme in 1916 and in the Ypres fighting of 1917 has been described elsewhere. Long afterwards he was awarded the D.S.O. for his consistently gallant and thorough work, but particularly for his superhuman efforts at Guillemont and Delville Wood. There was no more popular visitor at any of the company messes. He knew every officer, N.C.O., and man in the Battalion individually, and he was the friend of them all.

Major H. J. Wenyon took over command of the Battalion on the departure of Col. Whitty, and Capt. R. P. Baker, the adjutant, acted as 2nd-in-Command temporarily, 2nd Lieut. J. B. Cryer taking over the duties of adjutant.

In December, Brig.-Gen. W. F. Sweny, D.S.O., the popular commander of the 72nd Brigade, left for six months' duty in England, much to the regret of everyone in the Brigade. Brig.-Gen. R. W. Morgan, D.S.O., South Staffordshire Regiment, took over command, and remained with the Brigade until the end of the war. The early months of 1918 were marked by increased activity in the trenches, hard training, and digging of trenches in reserve battle zones, and a largely increased leave allotment. The last of these was popular, but there is some doubt about the others.

On January 1 the Battalion was inspected by

Gen. Morgan at Hancourt, and on the 2nd we moved to Vendelles. The system of holding the forward positions in depth was beginning to be adopted, and when we took over the line on January 5 from the 2nd Leinsters, we had one company only (B) in the front line, two in support, and one in reserve at "L.10 A." This was the map reference of the ration dump where the forward light railway started. Our transport used to come up to this spot each day.

The frost held until the 10th, and was so hard that no digging was possible, so instead we concentrated on carrying up enormous quantities of R.E. material to cope with the inevitable collapse of the trenches when the thaw should come.

In the war-diary for January 11 there appears a note: "Our trench mortars conducted a shoot on Ruby Wood." A word might be said about this wonderful wood. It was in the enemy front line a little on our left, and consisted of a number of straggling tall trees. Some said there were six—some seven. Some ingenious people used to report that the number changed from day to day. The Divisional Pioneer Battalion sent us up a serious message one day that they had distinctly seen one tree added to the number. Knowing the habits of the pioneer mess, we could only conclude they were gifted once more with second or double sight. Every tree in the wood was at one time or another accused of being artificial, made of tempered steel, and containing a battalion of Boche snipers and observers. The

Photo by] [*Thomas Fall, London*.

LIEUT.-COL. H. J. WENYON, D.S.O.
Commanding the Battalion December 1917—June 1919.

universal opinion of those in front was that it was just an ordinary battered old copse, and the Hun hated it as much as we were supposed to do. It was visited in 1919 with a view of photographing it and finally settling how many trees there were, but the wood had completely disappeared.

The thaw set in on the 11th, and the trenches began to tumble in a bit. Till the 13th we worked in every way with pumps and mud scoops to get things better, and certainly improved the trenches a lot. They were at least passable, and you could not get drowned in them when we handed over again to the Buffs on the 13th. We moved to camp at Montigny Farm.

Just before we came out it was decided that we would try a raid on a section of the Hun trench known by us as "Herbert's Post" next time up. Herbert was our name for the sentry they used to post there, or rather for the successive sentries, because we shot several of him. They never found out that we could see him from the chest upwards from a forward post of ours a little to the right, and with a telescopic rifle at 300 you could not miss.

D Company were to carry out the raid, and 2nd Lieuts. Carville, Crighton, and Janaway were to take part. While we were out of the line until the end of the month every man in the Battalion was out on working parties day or night. D Company were digging practice trenches for their raid, and later practising it. Other work was concentrated on the forward area, digging and wiring the intermediate (or Brown) line, various switch lines and defensive works and digging in the Bolsover Tunnel, a recent discovery of old underground works near Templeux. There were very few men available for work round the camp, but the few there were tarred and reroofed the huts, put in some stoves, and carried out many minor improvements.

On the 20th we had a practice stand-to in Hargicourt Trench. We were, of course, in the 5th Army at this time, and the activity on all sides was part of the strenuous efforts to form some sort of defensive positions in the enormous and threatened front held by Gen. Gough.

About this time the reorganisation of infantry brigades took place, and the 8th Battalion, The Queen's Regiment, left the 72nd Infantry Brigade and joined the 17th.

On the 28th there was a bombing raid on Montigny Camp. One bomb hit the new Theatre, "Daly's Theatre," where the Divisional Party had been giving a highly successful pantomime, "Aladdin," for some weeks. Several of the performers were hit, and three of our men were wounded.

On the 30th we moved up to the line again in relief of the 2nd Leinsters. The enemy was getting more alert in the front, and was beginning to snipe and to use his light trench mortars quite a lot. On the night of February 3rd/4th, D Company raided the enemy trenches. It was a silent raid, and gaps in the enemy wire had been cut previously by the 6-inch Newtons. The following is the official account of the raid, sent in at the time by Lieut.-Col. Wenyon :

" The raid on the enemy position at Herbert's Post was carried out last night.

" The assembling of the parties in No Man's Land was carried out in good order as arranged, and the circumstances seeming to be favourable, the Commander of each party took all his men to forward assembly positions on the enemy wire within a few yards of the Gaps. This was completed by the Right Party at 9.55 p.m., and by the Left Party at about 10.5 p.m.

" The Control Post had proceeded to its position, and Advanced Battalion Headquarters was in touch with 2nd Lieut. Janaway by telephone from 9 p.m. until about 11.10 p.m.

" At 9.55 p.m. 2nd Lieut. Carville led his party into the enemy trench. The wire had been perfectly cut, and the entry was made without difficulty. The party then proceeded along the trench northwards and reached a point where an enemy post is normally held. This was found to be unoccupied, and the party proceeded again towards a suspected dug-out. At this moment loud shouting was heard, and a large party of the enemy appeared on the high bank which forms the parados of the trench at this point. Our men were very heavily bombed from this bank, and further reinforcements in large numbers came up. One of the enemy, either officer or warrant officer, dropped into the front line and made off north. 2nd Lieut. Carville at once pursued him and fired four revolver shots into him at close range. 2nd Lieut. Carville did not consider it possible to search this man, as by this time the enemy party had been considerably reinforced, his party was hopelessly outnumbered and in great danger of being cut off.

"It was impossible to scale the parados, and the order was given for our party to throw all remaining bombs and return to their assembly positions in No Man's Land. One of the enemy stood up to throw a bomb, and was shot by 2nd Lieut. Carville.

"By this time 2nd Lieut. Carville had been twice wounded, but he succeeded in getting his party back to shell-holes outside the enemy wire. From here he ordered the party to return to our trenches in small parties, and finally returned himself at 10.45 p.m. The casualties in this party were several, but not serious. Unfortunately a trench mortar fell among a party of six entering our trenches in Curtain Lane and killed or wounded them all.

"The Left Party entered the enemy trenches at about 10.10 p.m., and the two blocking parties proceeded to their posts. Two of the enemy retreated down the communication trench, and were fired at by the blocking party.

"2nd Lieut. Crighton led his party to the 'T' head. This was found to be unoccupied, but very fully stocked with bombs. The alarm was given by two men running northwards behind the trench shouting 'English.' From the 'T' head our party made for the enemy trench, and just as they were entering it, 2nd

Lieut. Crighton was shot. Men appeared to come out of about four dug-out entrances behind the trench, as four patches of light could be seen, and there was a good deal of shouting. A word like 'Mull' was passed down from mouth to mouth, and our party was heavily bombed from the parados.

"Sergt. Vanner had taken command of the party, and all available bombs were thrown. The blocking parties co-operated. Sergt. Vanner then went to the blocking parties and gave the order to return through the enemy wire. He and Sergt. Allen stayed behind and attempted to get 2nd Lieut Crighton's body away. They succeeded in dragging him some fifteen yards towards the 'T' head, but they were seen and fired at by several rifles from about fifteen yards away. Another effort was made to move the body, but Sergt. Vanner was then heavily bombed as well as fired at. They could not get the body any further. Sergt. Allen was hit slightly.

"Outside the enemy wire two of our party were wounded. Sergt. Vanner and Sergt. Allen succeeded in getting them back to Bait Trench. They were all reported in at 11 p.m.

"2nd Lieut. Janaway was still at the Control Post. He succeeded in gaining touch with the Left Party, and kept me fully informed by telephone of their movements. He made four attempts to gain touch with the Right Party, but was prevented by heavy volleys of bombs and darts which were put down by the enemy in front of his own wire. The enemy was continually bombing his wire just in front of the Control Post, but 2nd Lieut Janaway and the sapper detailed for the work succeeded in exploding one Bangalore Torpedo in the wire.

"After this the enemy trench mortared the Control Post, very severely wounded one of the stretcher-bearers in 2nd Lieut. Janaway's party, and buried 2nd Lieut. Janaway. I then ordered 2nd Lieut. Janaway to return. With the help of Cpl. Andrews the wounded man got in with very great difficulty. 2nd Lieut. Janaway finally returned to our line at 12 midnight, having been at his post in very difficult circumstances, constantly being bombed and later trench mortared for three hours. His work in keeping touch with the Left Party and keeping me informed of what was happening was of the greatest value.

"It was quite impossible for either attacking party to get in touch with the enemy, and in both cases our party was considerably outnumbered. As far as possible, in the case of each of these parties, as well as at the Control Post, the plans made were carried out.

"The return to our trenches was carried out in orderly fashion. The total casualties in the raiding party and the trench garrison were:

"One officer and five other ranks killed.

"One officer and twelve other ranks wounded.

"Also two of the sappers attached were wounded. A large proportion of these casualties were caused through enemy trench mortars falling in one of the posts in our trenches and among the party of six men previously mentioned.

"Communication back and forward worked most excellently throughout. The wire out to 2nd Lieut. Janaway fortunately held until after the return of the Left Party."

It turned out that 2nd Lieut. Crighton was not killed, as he was subsequently reported wounded and prisoner of war.

The following decorations were awarded for this raid: 2nd Lieut. Carville and 2nd Lieut. Janaway, the M.C.; Cpl. Andrews, Bar to the D.C.M.; Sergt. Vanner, D.C.M.; and Sergt. Allen, M.M.

On the 5th we extended our Battalion front southwards. After a tour out at Hancourt, where the whole Battalion was working all the time on digging the Green Line (some five miles behind the front positions), and a tour up at Cote Wood, where we relieved Indian troops (the Sialkot Dismounted Cavalry Brigade), who had recently come up into the sector, we took over the Right Subsector from the 1st North Staffordshire Regiment. This was new ground.

Most of the front—which was a wide one—was held by isolated posts; there were no continuous trenches. We had an uneventful tour here. Battalion Headquarters were back again at our old friend The Egg—still very comfortable.

On the 22nd we were relieved and moved to Vendelles. A battalion mess had been built on quite luxurious lines there. Mess room, anteroom, officers' cubicles and bathroom were all complete. The whole place was made in a large Adrian hut, and had been very generously furnished out of the Divisional funds. News reached us here that we had finished our long spell in the Hargicourt Trenches, and we were going out for a rest. A draft of sixty-seven joined us at Vendelles during this tour there.

On the 27th we were relieved at Vendelles by the 2nd/7th Manchester Regiment (66th Division), and we had a most cordial handing and taking over. We marched away at 10 a.m., with the drums at our head blazing away at the tune of "Good-bye-ee, Good-bye-ee." We little knew how soon we were to be back in the district. The Battalion entrained at Roisel at noon, detrained at Villers Bretonneux, and marched to La Neuville near Corbie, where we were greeted by a deluge of rain. A large draft of 108 joined us on Roisel Station that morning.

We all had very pleasant memories of the Hargicourt Sector. Five months in the same district was a long time, and there was probably no place among our travels up and down the line which we had taken possession of and made our home quite so much as this one.

When we had settled down, we had some good times behind the line. Trips to Amiens were arranged fairly frequently, and the Division really assisted greatly in organising concerts and shows. The inauguration of the 72nd Brigade Concert Party, "The P.B.I.," came about at Vendelles in November, and started most successfully. Among the enormous working parties when we were out of the line there was little time for organised sport, but we had a few good football matches.

Even as we marched away from Vendelles that morning there was a feeling in the air that something was going to happen. We were indeed on the eve of tremendous events.

PHASE VII

WITH THE FIFTH ARMY AGAINST THE GREAT GERMAN OFFENSIVE

March–April, 1918

We detrained at Villers Brettoneux at about 6 o'clock in the evening of February 27, and marched through the rain to Corbie. There was some difficulty in getting into billets, and the position was not eased by the rain, which had now become a positive deluge. However, we were back in civilisation, and we were looking forward to a few weeks of it.

So much for our vain hopes. Soon after midday on the first day of our rest, rumour—which ever travelled faster than orders—indicated that everything was not right on the 5th Army front. During the afternoon we received orders that our long-expected rest was cancelled, and we were to entrain in the early hours of the next morning to take up a position in reserve behind Vermand. The disappointment of this turn of events may be imagined, and the situation is put before the reader here because it is one of those cruel turns of fortune which, apart from fighting and the trials of the trenches, made the lot of the infantry soldier a very hard one. The next morning back we all went into camp near Devise, a little village near Péronne.

We trained strenuously at Devise for a few days. We were now in the 19th Corps, commanded by Lieut.-Gen. Sir Herbert Watts, and were in reserve to the Dismounted Cavalry Division, who held the line in front of Vermand.

On Sunday, March 10, the 72nd Infantry Brigade were inspected by our new Corps Commander. Afterwards the whole Brigade marched past the General by platoons. One of our platoons (No. 4 of A Company) was placed as the smartest platoon on parade.

From time to time during our stay at Devise we reconnoitred the line we were shortly to take over. Finally, on March 11, we moved up to relieve troops of the 2nd Dismounted Cavalry Division at Vadencourt, and took the position of left forward battalion of the 72nd Brigade. Vadencourt was a beautiful place—once it must have been a superb château. All buildings were of course demolished, but the grounds remained practically untouched.

In view of the importance which later attached to the position at Vadencourt, it is necessary to give a brief description of the lie of the land and the method of holding the line.

The 72nd Brigade were on the right of the 24th Divisional front, and held a front of about 3,000 yards, with two battalions forward and one in reserve. The country was "rolling" from east to west. Vadencourt Château was in a commanding position on a ridge which practically formed our main line of resistance. A road ran due south from our Headquarters to

Maissemy, about a mile and a half from us, where was the headquarters of the Right Battalion. About 600 yards east of us was another ridge, on which and in the plain in front was the outpost zone to a depth of about 800 yards, and on a front of 1,500 yards. The Omignon River formed the right flank of our Battalion front, and ran at right angles to our position from Vermand, some 2,000 yards behind Vadencourt, into the enemy territory in front of us. The forward positions were held by isolated posts. There were no continuous trenches. There was a double line of trenches in the ridge in front of Battalion Headquarters already referred to, and Vadencourt itself was a large organised and well-sited redoubt, with trenches with good field of fire and machine-gun positions. The main road, known as Watling Street, from Vermand to Estrees, was our left flank. Almost in the centre of our outpost zone was a huge mound known as The Tumulus, standing some fifty feet high. Such registration as the German gunners did was usually carried out on this mound.

Most of our positions were moderately well wired, and had a good field of fire. There was a very wide No Man's Land—about 1,000 yards. At Vadencourt itself, in addition to the flank defence formed by the Omignon, there was strong and well-sited wire, and on our right front an arm of the river formed a swamp at the bottom of the valley about 300 yards from our fire positions. From the point of view of stationary warfare, this seemed to be an ideal sector of

THE TUMULUS.

VADENCOURT CHÂTEAU.
Battalion Headquarters was near the two trees on the right.

the line. There was little shelling, and it was a case of lying dogo simply during daylight, and thoroughly beating up No Man's Land during the night. Large patrols were out each night from dusk till dawn, about eighty men and four officers being employed.

On the whole all the positions except one or two of the foremost were comfortable, the weather was fine; and when we first had a good look at the delightful countryside, it was almost impossible to think of anything but a peaceful life there. Very soon, however, it was to become an inferno of every kind of beastliness in the way of shot and shell. We knew we were brought back into the line for a reason, and work on the improvement of defences at all possible times, the constant instructions of all ranks as to the exact importance of the various positions, and constant intelligence reports of activity extraordinary behind the enemy lines made us feel in our bones that the charm of our present outlook might not last for ever. We started then on this evening of March 11 for a long tour in this position. Two companies were in the outpost zone in front; two companies and Battalion Headquarters in the Vadencourt defences. Inter-company reliefs were carried out every four days, and we began to settle down to routine work.

In spite of the seeming uncertainty of things, leave was never better in the whole war. When the great enemy offensive commenced on March 21, about 200 officers and other ranks of the Battalion were on leave to England. On March 19 B Company were holding the outpost zone on the right, D Company on the left. A and C Companies were at Vadencourt Château with Battalion Headquarters. On our right the 1st North Staffordshire Regiment were in the front line with Headquarters at Maissemy.

On our left were the 3rd Rifle Brigade (17th Brigade).

On March 20, which was very quiet, the adjutant, Capt. Baker, went on leave. In the evening report was received at Battalion Headquarters that according to information given by a German prisoner, an enemy attack might take place on March 20 or 21. The usual patrolling was carried out on the night of the 20th/21st, and no abnormal enemy activity was reported. The night was clear and fine, but just before dawn a thick mist began to develop which shortly became so dense that it was impossible to see more than fifteen yards. It was in this famous mist that the Germans commenced their grand offensive against the British 3rd and 5th Armies, which was designed to cut between the British and French Forces and win the war in a few weeks.

At 4.30 a.m. on March 21 a general bombardment by the enemy commenced, and we were in the thick of it.

Vadencourt Château was very heavily shelled for six hours, largely with gas shells. The outpost zone and trenches on the ridge were being knocked to pieces.

A very gallant young officer of C Company, 2nd Lieut. C. C. Baring, unfortunately got badly gassed by almost the first gas shell thrown over, and later died of the effects. Apart from this there was no known case of a gas casualty in the Battalion. About 7 o'clock Lieut. G. S. Bowen, the Intelligence Officer, set out to visit the forward companies. He got safely through a perilous journey, and returned to Battalion Headquarters about 9 o'clock. The trenches in which were the Company Headquarters on the ridge in front had been badly knocked in in parts, but no attack had developed in this area.

The enemy gunnery had certainly been marvellous. In the previous weeks there had been practically no registration by their artillery, and almost all shooting on March 21 must have been by map or by sound ranging. Yet there was scarcely a battery behind our front which did not lose half its guns, and trenches and redoubts were shot up with wonderful precision. Just behind Vadencourt were two 4·5-inch howitzers, which had been in position two days, and fired a few rounds only.

Both these guns were knocked out by direct hits very early.

About 10.30 a.m. the order was given for gas masks to be removed, and a few minutes later the first report of the enemy advance came in. It was still too misty to see more than a few yards, and scouts had therefore been sent out on the flanks. One of these from the road to Maissemy reported that the enemy had attacked the North Staffordshire Regiment from the right flank and rear, and had overwhelmed them and was now advancing on us straight up the Maissemy road. How much of this was true it was impossible to say, but it was soon obvious that something serious was wrong, for machine-gun and rifle bullets began to whistle among us from the direction of Maissemy on our right, and slightly from the rear. The bridge over the Omignon was demolished at once, and just in time, by Lieut. Christison, R.E., of the 103rd Field Company, in the face of the enemy. This bridge was the only crossing of the river anywhere near our position, and we now had a fair protection on our right flank against direct assault, but not against enfilade fire, which was very troublesome all day long.

About 11.30 a.m. the fog began to clear and the sun came out.

At this point it may be well to mention the officers and warrant officers who held the commands within the Battalion :

Commanding Officer, Lieut.-Col. H. J. Wenyon.

Adjutant, 2nd Lieut. J. B. Cryer.

Intelligence Officer, Lieut. G. S. Bowen.

Medical Officer, Capt. W. A. Kellogg (U.S. Army).

Company Officers :

B Company (Right Forward Company), Capt. C. R. H. Allworth.

D Company (Left Forward Company), Capt. F. C. Needham.

WATLING STREET, COOKER QUARRY, THE SUPPORT POSITIONS, AND IN THE DISTANCE THE TUMULUS.

THE BRIDGE OVER THE OMIGNON RIVER.

A Company (at Vadencourt), Capt. W. J. Ewen.

C Company (at Vadencourt), Capt. S. G. Thompson.

Regimental - Sergeant - Major, Regt.-Sergt.-Maj. R. Rankin.

Transport Officer, Capt. J. N. Arnaud.

A/Quartermaster, Lieut. H. E. Outram.

Regimental-Quartermaster Sergeant, Regt.-Quartermaster-Sergt. R. Ifould.

A number of officers, including the Adjutant, Capt. Baker, the Quartermaster, Capt. J. R. Wood, and the Officer Commanding D Company, Capt. F. Proctor, were on leave.

As visibility became better at 11.30 a.m., it was possible to discover at last something of what had happened.

The enemy were seen to be occupying Mareval and Pontru Trenches on the ridge in front of Vadencourt. Capt. Needham, Officer Commanding D Company, got back wounded to Battalion Headquarters. He reported that after very heavy shelling, his company had been attacked from the right flank and from the rear, and very few could have escaped. Shortly after this two prisoners were brought back by one of B Company's runners, and a few men of B Company (mostly wounded) came to the dressing station. The following account of the fate of B Company is given by Capt. C. R. H. Allworth:

"At 4.30 a.m. on March 21 the enemy opened a very heavy bombardment on our positions, chiefly on the village of Pontru and on the forward outpost positions; a great number of gas shells were used, together with the high explosive. By 5 o'clock the telephone wires between Company Headquarters and the forward platoons and Battalion Headquarters had been cut by shell-fire. During the bombardment the Battalion Intelligence Officer and a runner visited me; they stayed for about ten minutes, and then returned. I then decided to go up to visit the forward platoons; my runner accompanied me. We had left the trench, and had proceeded about 200 yards, when two Germans appeared out of the mist, one was a wounded man and the other a medical orderly or a stretcher-bearer; the latter escaped in the mist, but we captured the man and brought him back to Company Headquarters. He could speak neither French nor English, and since I could not speak German, he was not of much use to us. This was our first indication that the enemy had attacked.

"The enemy had occupied Pontru village and all the foremost positions. About 8.30 a.m. they attacked our line (Pontru Trench and Mareval Trench). The wire in front of the trench had been untouched by the bombardment, and owing to our Lewis-gun and rifle fire they failed to get through it. The enemy then brought up a machine gun on the flank bridge across the Omignon River on the right of our sector; this was put out of action at once by our right Lewis gun. They then brought up another gun, and, owing to a jam in our Lewis gun, put our team out of action. A number of the enemy came down the banks of the stream and rushed the post and occupied the right of Pontru Trench. We attempted to bomb them out, but they got round in the rear of the trench, and we had a rather exciting retreat down the trench to Company Headquarters, where we established a block.

" By this time the enemy had broken through on the left, and were coming down the Bellenglise–Vermand road. I ordered a few men to hold the " block," and the remainder I sent up the trench to convenient positions to fire at the enemy on the road. The Germans suffered very heavy casualties on the road ; a machine gun from Cooker Quarry did some splendid shooting. After about three-quarters of an hour of this, which under different conditions would have been very good sport, we ran out of ammunition. The enemy had been rifle grenading us and sniping us very heavily, and only very few men remained. We could see the Germans digging our support line, and there were great numbers attacking our rear positions. We then, with fixed bayonet, got out of the trench and attempted to get back to the Battalion's main positions, but we ran into large numbers of Germans and were overpowered.

" The men of my company behaved splendidly. Though several of them were wounded, they carried on firing, and the enemy suffered very heavily."

Had we not known it from the weight and length of the bombardment, it was obvious, as soon as we could see, that this was no ordinary attack. The effrontery of the enemy was astounding. He was marking out the line of his advance with white flags, and far in the rear the forward movement of battalions with full transport could be seen in all directions. Message after message was sent down asking for artillery fire to be brought to bear on these wonderful targets, but our guns were practically all silenced, and not till late in the day did any begin to speak at all, and these were shooting very short into our own positions. The general advance of the enemy in large numbers towards Vadencourt and Cooker Quarry (held by the Rifle Brigade on our left) began about 12 noon. Vadencourt was subjected to very heavy machine-gun fire from three directions. Finding the obstacle of the river and swamp on our right front too difficult, the enemy attempted to advance across our front towards Cooker Quarry. He was in close formation about a battalion strong. Our Vickers guns, Lewis guns, and rifles did tremendous execution, and very few of the enemy got back. This success braced us up considerably, and we felt there was yet a chance of holding on to our position if only our flanks could hold. No information came through to us as to the general position, even on our own divisional front.

In the afternoon the Rifle Brigade were still holding a line level with ours, but on the right the enemy was established in Maissemy and had advanced about a mile beyond it, and was trying to get machine-gun posts on the river bank.

The scene on our front during the afternoon was most exasperating. Several frontal attacks by the enemy were frustrated by our rifle fire, but the steady movement of enemy forces from the rear could not be checked. All we could do was to engage them with Vickers guns at 2,000 yards' range. Enemy aeroplanes swarmed overhead and bombed us at intervals. An observation post had been established on The Tumulus, and a group of mounted men remained there all the afternoon.

In the early evening the enemy commenced to shell Vadencourt and its surroundings heavily again; trench mortars had been brought up, and things began to get very unpleasant. In spite of the precarious nature of our line, it appeared that we could hold on if the troops on our left in Cooker Quarry held. We had plenty of ammunition, and although we were reduced in numbers and several machine guns had been knocked out, we could still bring a very withering fire to bear on anything like a frontal attack.

At 7.15 p.m. the enemy commenced to advance towards Cooker Quarry, and unfortunately both quarry and trench were captured, the garrison, some 200 strong, retiring in the direction of Vermand. The enemy immediately commenced to advance in large numbers up Watling Street, and towards the high ground on our left flank. This left us with both flanks hopelessly in the air. We engaged the enemy with every available rifle, Lewis gun, and machine gun, and considerably delayed his advance. About this time we received the only written message of the day from Brigade Headquarters, sent off about five hours before. This informed us that a counter-attack to retake Maissemy under an artillery barrage would take place at 7.30. In view of this we decided to hold on. This message, as we learned afterwards, was cancelled by another one which we never received, and the counter-attack did not take place. We were now truly in "Queer Street." Every minute we were expecting the bombardment and attack on Maissemy to commence, every minute also we were expecting an overwhelming attack from our left flank. The outposts of A Company on this flank did splendidly during these tense minutes. They harassed the enemy furiously, and undoubtedly deceived him as to our actual strength at this particular spot. It was barely a platoon, but it might well have been half a battalion. However, it was obvious this could not go on for long. We were badly placed to meet an attack from this flank, and as soon as the enemy discovered the real position he could overwhelm us by probably twenty to one. Still there was the hope of the Maissemy attack, and we held on.

The next move by the enemy was a bombardment by howitzers and trench mortars, added to a rattling and continuous machine-gun fire on this unfortunate left flank of ours. Tyler and Malby, two young officers of A Company, were in charge up there, and they and their men stuck it splendidly. But they were losing men steadily, and something had to be done at once.

At 8.30 p.m. it was decided that we would make an attempt to withdraw under cover of darkness to the Brown Line at Bihecourt some 600 yards behind. This was no easy matter, and it was essential that we got away unknown to the enemy. A strong rearguard was organised, and took up positions on the threatened flank.

Lieut. Goulden, 2nd Lieut. Pfeuffar, and 2nd Lieut. Tiley, with a number of men with four Lewis guns formed the guard, and they set up a powerful demonstration of fire for nearly an hour. There was no panic in the Battalion whatever. A definite order of march was laid down, and a section at a time the whole garrison withdrew to Bihecourt.

The rearguard got back about 10 p.m., and, thanks to their work, we had achieved almost a miracle in getting back without a single casualty. Two companies of the 13th Middlesex reinforced us in this line, and the whole force was properly organised from the river northwards, and into touch with the 17th Brigade by 12 midnight.

So ended March 21, 1918, one of the most memorable days in the whole history of the Battalion. The brief sketch of the main events of the day gives a very feeble picture of the actual fight. Countless acts of heroism could be recorded if space allowed. The whole Battalion fought splendidly. Vadencourt could well have been described as hell during almost the whole of that day, yet no one ever thought of wavering, and again and again the enemy was thrown back with heavy losses. It was not till afterwards that we learnt with pride that the holding of the ridge north of the Omignon River had such an important tactical bearing on the general situation, and that the 24th Division was the first division in the British Army to be mentioned by the Commander-in-Chief in his first dispatch on the great fight against the German offensive.

Congratulations were showered on the Commanding Officer for the work of the Battalion. A message had been sent to us ordering us to withdraw to Bihecourt in the early afternoon. The message had never arrived, but the Brigade Staff, seeing that we did not come back, had given us all up for lost. A council of war took place in an old cellar in Vermand at midnight. Everything was fairly quiet now, but it was obvious that by the morning the enemy would be ready for another great smash. So far he had achieved very little success on our front, but our numbers were sadly depleted, and we had seen almost countless numbers of men, horses, guns, and trench mortars being hurried into the forward positions obviously for the next day's work. It was decided that a stand should be made on the Brown Line, and the 72nd Brigade was hastily reorganised.

Col. Le Fleming of the 9th East Surreys had been killed in the afternoon; Col. Pope of the 1st North Staffordshire Regiment severely wounded in the morning.

Col. Anderson, who came up to reinforce with

dismounted cavalry, joined the East Surreys, and took command of the force south of the Omignon. Col. Wenyon was given command north of the river.

A certain amount of fitful sleep came to most of us during the night, and food was taken very early in the morning. There was another very dense fog at dawn, and the enemy did not begin his advance so early. About 9 o'clock he commenced a hurricane bombardment of Vermand and the Brown Line area, and immediately launched his attack on Bihecourt. Possibly the cases of excellent port wine which we had been compelled to leave behind at Vadencourt had made him reckless, and in spite of casualties the enemy advanced almost right up to our trenches. Fortunately there was some wire in front of this line, and most of the attacking force that had got so far died a few yards only in front of the muzzles of our rifles. Here and there over the 300 yards of open country in front of us small pockets of the enemy established themselves in shell-holes and old trenches. A supply of rifle grenades which were found in the Brown Line were used with great effect by C Company in driving away these enemy posts. The enemy's trench mortaring was again very unpleasant. The manner in which he had moved up these great engines was almost uncanny.

Constantly during the morning the attacks were renewed, but he met the same steady front he had tasted the day before. The two companies of the 13th Middlesex took up the fight splendidly, and we can really be said to have taken tea with the Hun on this morning. Several times he reached our wire in small numbers, but he never got into our trenches. From the noise on our left we were under the strong impression, however, that all was not well there. It was pretty clear that the enemy had fairly penetrated, but it was not until much later in the day that we knew that during the morning he was fighting through Bernes, some five miles behind us.

About 1 p.m. the enemy succeeded in driving in our troops at Villecholles, south of the river, and reached the outskirts of Vermand. So suddenly did this happen, that it seemed that the only way back for our men was going to be closed before we could move. An immediate order to withdraw was given, but it was not until 2 p.m. that we actually left the Brown Line. By all the rules of warfare, none of us should have got back alive. Our only way back was along a dead straight road (our own wire made it impossible to go across country). This road was being shot along and across by a murderous fire from well-placed machine guns. We certainly had many casualties on the Vermand road, but the miracle was that so many did get safely back.

We got on to the road from Vermand to Monchy Lagache, and it was a case of a straight march in small parties. The enemy was through a long way on our left, and too close to our flank to be pleasant. He was through even further on our right, but beyond the river, and not so close. We were peppered with machine guns along the road, but we got through, and we were ready for a rest and some food when the Brigade concentrated in some old trenches at Monchy at 8 o'clock in the evening. After these two days of battle we realised the tactics, now so well known, which the enemy was using for his great offensive. The battling all along the line was calculated by him to discover some weak spots in our resistance. Immediately he had penetrated anywhere, he sent reserve mobile troops with motor machine-gun sections if possible through the gap, and very often succeeded in cutting off troops which had resisted. This is what happened to part of our front companies on the first day. The enemy had penetrated right through on their right, and, crossing the river between Vadencourt and their trenches under cover of fog, he had attacked them simultaneously from the flank and the rear. So it was all the way back on this sustained rearguard fight. Every day we were fighting the Hun very hard, and at close quarters; at the end of the day we found our positions intact, and were ready to take up the fight with him again at dawn. Then, just as we were hoping to snatch an hour's rest, we were ordered back to conform with a new line for the next day's battle, the enemy having succeeded in getting through some miles on one of our flanks.

We passed through the Green Line, held on this front by the 50th Division, who had been hurried up from rest in motor-buses. The position did not inspire us with much confidence as we passed through it. On most of the front the trenches were only a few inches deep, and the front held by the 50th Division must have been a wide one, for the troops were very scattered. However, the trenches were well sited and the wire was quite good. One party of our men, under the Regimental-Sergeant-Major, stayed and held a post in the Green Line until nightfall, when they joined us at Monchy Lagache.

We found the transport under Capt. Arnaud at Monchy, and heard the first news of their experiences. The transport lines had been fairly blown up at Vermand on the 21st. Cpl. Verrall and one or two men, directly the bombardment started, got the animals out and stampeded them on the road behind Vermand. Here they were all collected, and it was a miracle that we did not lose a single horse. Had they remained in their lines for half an hour, most of them must have been killed. The Brigade Transport Officer had taken the transport right back to Monchy straight away, and in consequence there were no vehicles available forward to remove our Quartermasters' stores from Vermand, most of which had to be abandoned. The cookers were going strong at Monchy, and we had a very welcome hot meal before getting our first rest for forty-eight hours.

All night the hangars at the great aerodrome at Monchy were burning, and lit up the country for miles. The enemy shelled the village in a random way, and the night was bitterly cold.

At dawn breakfast and as fine a tot of rum as was ever drunk was issued. The Commanding Officer was called to a council of war at Brigade Headquarters in the village.

Col. Wyatt, who was commanding the Brigade in the absence of Gen. Morgan, set out the plan for the day. The Green Line had apparently been fairly easily forced by the enemy, and our Division was ordered to withdraw across the Somme at Falvy. The 24th Divisional Reinforcement Camp had been formed into a battalion, and were attached to the 72nd Brigade for the day. Our brigade was detailed to form the rearguard on the eastern bank of the Somme for the rest of the Division to pass through. The other brigades were in the trenches on our right, and we were to stop at Monchy till they had all moved off. It was the third foggy morning, and this seemed all to the good for our particular job. However, it seemed the orders had not reached the other brigades so soon, and we had to wait for them to move. All orders were issued and the order and manner of march laid down, but it was after 10 o'clock before we started.

Then, just as we were leaving Monchy to take the track across the three miles of plain to Falvy, our Commanding Officer, who was in command of the rearguard, found the Rifle Brigade digging themselves in behind the village. Gen. Stone, commanding 17th Infantry Brigade, was in attendance, and he was of opinion that one battalion ought to be left there for a time to prevent the enemy if he took the village from debouching on to the plain. The fog was just beginning to disappear, and it appeared that this plan was unwise in the circumstances. However, Gen. Stone wished the Rifle Brigade to fight this position, and he asked for one company of the East Surreys to scout through the wood as a flank guard on his left. The company was accordingly detached, and worked its way through the wood. Later in the day this company got to the Somme, when it was a perfect inferno, and at a place where there was no bridge. Many of the men were lost, and the remnants had to swim the river.

The remainder of the Brigade got to a position on the east bank of the Somme, as ordered, and by this time the whole of the Division could have been safely over the river. About 11 o'clock the sun came out, and every blade of grass on this wide plain was visible. The whole Division, including attached troops, were there. When the Rifle Brigade began to come back at about 11.30 they were shelled unmercifully, so were the Reinforcement Camp Battalion just behind them, and so were the rest of the Division on the bank of or crossing the river.

Our cheery little friend, the Brigade Signalling Officer, Lieut. G. F. Hopkinson, did marvellous work during the terrible time that followed. Again and again he went up on his motor-bicycle across the plain and brought back wounded men to the dressing station across the river.

After every other unit of the Division had crossed the river the 72nd Brigade crossed at about 3.30. It had been a horrible day, and we are firmly convinced that had the whole division marched away from Monchy in the fog, the retirement would have been completed without a casualty.

Just before we crossed the Somme a squadron of cavalry rode over the bridge eastwards. They could do very little at this stage, but they seemed to be spoiling for a fight, and were acting under orders given several hours earlier. With the greatest gallantry they attempted to round up the machine-gun nests which the Germans had now pushed forward, but they suffered very heavy casualties both to horses and men, and had to return. A battalion of the 8th Division passed us as we crossed the river. They were to act as an outpost to the main line on the west side of the river. The division had been sent up from a rest area to hold the line of the Somme, and the men were in splendid fettle as they swung across the bridge.

So with wishes of good luck on both sides we passed through the 8th Division to the village of Licourt, some four miles off. We could not have gone to a more unfortunate place. It was peaceful enough, and our Battalion was got together and camped under the sky in a big field. But in the village was the headquarters of an Infantry Brigade of the 8th Division.

The Brigadier-General commanding this brigade was in doubt about his positions on the river. He found Col. Wyatt, who was commanding our Brigade, and our Commanding Officer was summoned to the Brigade Headquarters. The General stated that he had not received information of the arrival of one of his battalions. Our Commanding Officer pointed out that this battalion had passed through in the direction of the river in buses some time before. This would not do for the General. He ordered *us* to go.

Our men had been at it fighting or footing it for twelve hours, after two days of hell. They were very nearly dead beat, and now they were ordered back to the crossing of the river they had just passed, almost certainly on a wild goose chase. The Colonel decided to go down to the place himself first, and orders were left with the Adjutant that the Battalion should move off half an hour later.

Arrived at the river, he left his horse and groom at an old church and went on. He found, as expected, the Battalion was in position, and had been on the spot for some hours. It was now getting dusk and he could not find the horses, so he came back on foot, but unfortunately by a road parallel to the one down which the Battalion was marching. The Battalion had to proceed right to the river, and were then sent back. They arrived back at Licourt about midnight, and shortly afterwards Capt. Arnaud brought up the cookers and a wonderful hot meal. The night was again very cold indeed, and it was a case of lying down in the open. This meal was the finest thing we had seen for a long time. In spite of the cold and soreness of limb, we slept like logs,

On the next day, Sunday, March 24, the 72nd Brigade was ordered back early in the morning to Chaulnes, some six miles back. There were welcome signs of our own artillery on this march back. Big guns were in position, and speaking to some purpose on all sides, and more were lumbering up as we marched. At this time we were supposed to be marching out to be relieved by French troops, but we did not really believe it. We had heard that there had been some hitch about demolishing some of the Somme bridges, and the river positions did not seem to be altogether secure.

We arrived at Chaulnes about 9 a.m., and as we were now some miles at any rate from the enemy, we had a few hours' relaxation. The feet of every man in the Battalion were attended to first, and much-needed water and soap were brought into use. Kellogg, the Medical Officer, had a busy morning.

At noon a council of war was summoned hastily at Brigade Headquarters. The enemy was reported to have forced a crossing of the Somme at Béthencourt, and a gap had been formed between the 8th and 20th Divisions. The 72nd Brigade was ordered to march to Fonches to take up a position there. The American officer in command of a big dump, which had been set on fire, met us just before we left Chaulnes with the comforting news that the enemy was in sight and that we would never reach Fonches. As we marched across the railway, some excited French Engineers were trying to blow up the permanent way, and were doing their best to bring it off as we were crossing. In view of the news we marched to Fonches by companies, with strong flank guards and advanced guards thrown well out. We arrived there without incident at 3 p.m., and found Fonches was a place of perfect peace. The Brigade was disposed in trenches defending the Hattencourt –Fonches–Fonchette road, the 9th East Surrey Regiment on the right, the 8th Royal West Kent Regiment on the left. The 73rd Brigade were on our left in front of Punchy, and the 17th Brigade on their left in front of Omiécourt. The Battalion's dispositions were quickly made, and close liaison established with the battalions on our right and left. Battalion Headquarters were established at Fonches Church. Very soon it was clear that the enemy was not within miles of us yet, and the cookers, which had been kept a little behind, were ordered up. Fonches was a pretty little village, with an old village green; the weather was delightful, and we wonder if anyone who was there will ever forget the wonderful picnic on the green which we had on the arrival of the cookers.

We felt much fitter after this, and every man was given a task in strengthening our trench positions. The work went on right through the night, while joint patrols of our men and the 7th Battalion Northamptonshire Regiment were out under 2nd Lieut. Pfeuffar, through the dense wood and along the valley east of Fonches as far as Curchy. By daybreak we had greatly strengthened our positions along the road.

About 7 a.m. it was reported that the 8th Division were to make an attack in the direction of Morchain and Béthencourt, in conjunction with French troops. The 24th Division was ordered to move up in support of the attack, and the 17th and 73rd Brigades took up a position on the line Dreslincourt to Pertain. The 72nd Brigade took up a position in support, astride the railway west of Hyencourt-le-Petit. This meant a march of about two miles north from Fonches. The Royal West Kent Regiment deployed north of the railway in three lines, D and C Companies and the North Staffordshire Regiment (the remnants of whom were now attached to us) in front, A and B Companies in support, Headquarters in reserve. The enemy shelling became very unpleasant—the 17th and 73rd Brigades were driven in, and at about 1 p.m. the 72nd Brigade was ordered to return to Fonches and hold the line Hattencourt–Fonchette to the main cross-roads 800 yards north-east of Fonchette. After some difficulty—the main road being now the mark for enemy guns—we took up the new positions, and the Commanding Officer inspected the left flank of our position. It was found that the 73rd Brigade front was at right angles to ours, and running back towards Hallu. We had a telephone line to Brigade from Fonches Church, and the position was immediately explained to General Morgan, who had just got back from leave full of martial ardour.

Our Colonel was ordered to do all he could to re-establish this flank. *In any case we were to hold out at Fonches to the last.* The 9th East Surreys were in position on our right, and, leaving Maj. Clark in charge, Col. Wenyon went with his servant (Pte. Smith) to see what could be done on the left. The enemy had advanced considerably opposite the 73rd Brigade front, and appeared to be occupying Hyencourt-le-Petit, and positions south of that village. Heavy machine-gun fire was being directed on the area north and west of the cross-roads which formed our left flank. After about two hours, two battalions of the 73rd Brigade were got forward to the belt of wire running north and south just west of the Fonchette–Omiécourt road. Our position being now secure, Col. Wenyon returned to Fonches.

The Battalion had been having a bad time. The Fonches cross-roads was, and had been for some time, the mark for most things the enemy could throw over. The spot was under particularly heavy machine-gun fire, and we had our first experience of being shot up by field guns at point-blank range. Capt. Ewen, commanding A Company, had been killed, the Adjutant, 2nd Lieut. J. B. Cryer, had been wounded, and we had had many other casualties.

Kellogg and the medical officer of the East Surreys had their dressing station in the village road by the green, where we had picnicked so peacefully the day before, and they were attending to wounded without ceasing all the afternoon. The enemy was desperately close to this horrible cross-roads, and once again we could see hordes of men coming up from the rear, ready to reinforce his forward battalions. Still our garrison was well disposed to meet attack, thanks largely to the hard work of the night before, and the officers and men were ready to fight the position to the last. We had an exciting evening, and thrust back attack after attack. All the afternoon the enemy had been collecting in large numbers in the wood just south of this valley. After many messages had been sent down, at last some real heavies of ours got on to this wood and did excellent work for some hours.

Shortly after 8 p.m. determined efforts were made by the enemy to break through our lines. A party succeeded in getting into a disused trench about twenty yards in front of A Company. A cutting-out movement was at once organised, and after a stiff fight, in which 2nd Lieut. Stevens was wounded, the enemy retired, leaving behind a store of bombs and ammunition. We then put a post in this disused trench and established a block there.

The night was clear and moonlight, and, apart from isolated sallies by the enemy, which were met by our fire and dispersed, nothing of note occurred. In one of these sallies we secured a hefty Prussian prisoner.

The action of our Padre, Rev. G. D. Barlow-Poole, might be mentioned here. Having heard at Brigade Headquarters the orders that we were to fight Fonches to the last, he deliberately came up and spent the whole time with us.

Knowing what our orders were, 2nd Lieut. Cryer, though shot through the arm and in considerable pain, had refused to go to the Field Ambulance, and behaved with the greatest gallantry.

Numberless acts of splendid heroism could be mentioned, but space does not permit. To mention one would be to mention every man. Throughout the companies the word had passed, " When it comes to it, three Huns for each of us," and every man was determined to carry this out. Men were not recklessly throwing away their lives, but wherever the fight became hottest, or when any point was gravely threatened, there was a spontaneity about the measures to put things right which made our attacks irresistible. In the midst of all this noise and fury Capt. Arnaud succeeded in getting a cooker up near enough after dark for us to carry hot soup and stew, if you please, to every man in the Battalion.

The cross-roads at Fonches will surely always live in the minds of the little garrison who hung on by the teeth there through that awful day.

The end of it was that at midnight we had orders from Brigade to withdraw to positions at Hallu. We could hardly believe it, but the orders were clear enough. The enemy was very close, and the withdrawal had to be done with great care. It took a long time creeping, section by section, along the grass border of the village road,

FONCHES.
The Cross Roads.

WARVILLERS.
Some of the billets occupied by the Battalion.

out beyond the village, until it was safe to march in column of route on the road itself. The withdrawal was made in perfect order, without casualties, and we took up positions at Hallu by 3.30 a.m. About 5 a.m. it was possible to inspect the new line, and a gap was found between ourselves and the 9th East Surreys on our right. A Company, now commanded by Capt. J. H. Selfe, moved up from support to fill the gap.

We had had practically no rest, and the mental strain of the day before, added to the cold and physical weariness, left us particularly limp at dawn on March 26. We got a hurried breakfast, which improved matters considerably, and it was clear that we were soon to be at it again. At 6 a.m. the empty village of Fonches was being shelled to extinction, and the positions we were in at Hallu did not inspire us with a great deal of hope.

Col. Wenyon was ordered to take up Headquarters at Hallu in a dug-out with a direct telephone to Divisional Headquarters. We managed by this means to get some artillery fire to bear on the advancing Hun, but the shooting was wild, and many shells fell in our trenches. At 8 a.m. we received information from Brigade that the enemy had broken through the entrenching battalion on the right of the 9th East Surreys, and we were to withdraw fighting in echelon from the right. Capt. Thompson, who was commanding in our forward trenches, was accordingly ordered to conform with other units. Shortly after this the left broke also, and the 73rd Brigade withdrew. This left the remnants of the 72nd Brigade with both flanks in the air once more. The advanced enemy were within 200 yards of our trenches, and were advancing in vast numbers. Capt. Thompson informed the East Surreys that he proposed to withdraw by platoons from the left. The enemy was now very near our trenches, but was being well held by our fire. An orderly retirement commenced. Maj. Clark, with part of the East Surrey Regiment, still hung on to his position, and their battalion headquarters and a considerable proportion of their men were overwhelmed and captured.

We can well imagine Clark, dogged old soldier that he was, hanging on like grim death to that bit of trench. It was not till long afterwards that we heard he was wounded and a prisoner, and we all missed him greatly in the Brigade for the rest of the war. There can be no question of the gallantry of himself and his officers and men.

After the Battalion left Hallu, they were harassed from all sides by machine-gun fire, and some half a mile back they were stopped by a senior officer, who attempted to form a line in shell-holes. It seemed that he on this morning had more or less lost his reason, and by his action he placed a large body of men in great danger of capture. Capt. Thompson finally withdrew, and most of the men succeeded in getting back.

The Division now concentrated for the defence

of Rouvroy. Our men were utterly exhausted, and were kept in reserve at Warvillers, where they got a clean up and a night's rest. Here the Battalion was reorganised, and from now onwards was composed of the remnants of the whole brigade. A number of new Lewis guns were issued to the Battalion on that day. The 72nd Brigade now consisted of the Composite Battalion under Col. Wenyon, the 15th and 19th Entrenching Battalions, and a battalion formed from the officers and men of the 18th Corps School.

After close observation by low-flying enemy aeroplanes, the battle for Rouvroy and a Copse known as "Eight Trees" commenced about 10 a.m. on March 27, and continued throughout the day. C Company were disposed in trenches north-east of Rouvroy; A Company and the North Staffordshires in the outskirts of Rouvroy. The East Surrey remnant were with Battalion Headquarters in Warvillers. After ding-dong fighting all the morning, the enemy took Rouvroy from the Entrenching Battalion at about 2 p.m., the locality about "Eight Trees" having been forced by the enemy some time earlier.

Dispositions were then made for the defence of Warvillers after the North Staffordshires had made an attempt to re-establish the position in the south-west part of Rouvroy with the bayonet. A strong line was selected defending the villages of Beaufort, Warvillers, and Vrély. In this line were the two Entrenching Battalions and the 18th Corps School. After dark the

Composite Battalion was withdrawn into reserve in Warvillers.

Throughout the afternoon and evening parties of the enemy frequently attempted to debouch from the south-west of Rouvroy towards the trenches held by A Company and the North Staffordshires. Many casualties were inflicted on the enemy by our fire, and any serious attempt to advance was stopped. The new dispositions were completed early in the morning of March 28, the 72nd Brigade, reinforced by the Machine-gun Battalion, being in the line defending Warvillers, the 73rd Brigade being on our left, and the 17th Brigade in front of Vrély on their left. The enemy attacked at 8.30 a.m. in large numbers. The line on our Brigade front held firm, and great loss was inflicted on the enemy by rifle and machine-gun fire. In front of Vrély, however, our troops were beaten back, and on the right south of Beaufort a break also occurred. Warvillers had been shelled very heavily during the infantry attack, but we held on to the last possible moment. When the order was finally given to withdraw, there were a number of the enemy established in the village with us. We had several men sniped at very close quarters, a German post being in one corner of the Château grounds before we left.

The company commander in the front line, when he received the message to withdraw, did so reluctantly after manning his parapet all along the line, and giving the enemy ten rounds rapid fire.

Altogether the Hun had a pretty bad time in this attack on Warvillers.

The Division moved from here to the line between the villages of Caix and Le Quesnel, which was manned by about 1 p.m. Masses of artillery, including the first groups of French 75s we had seen were congregated just behind this line. The artillery horses had had a bad time from aeroplane bombs shortly before we arrived. The position was a good one, and we had a splendid view of the country and a good field of fire from some old trenches. But there were ominous-looking woods some 600 yards in front, and our flanks were very heavily wooded. It took some time for the enemy to move from Warvillers on to our front, but by about 2 o'clock he had got machine-gun posts in the edge of the wood in front, and one or two in some isolated orchards nearer to us.

Our ammunition was, as may be imagined, very limited, and we could not afford to do any continuous shooting at long range. Not so the enemy, who kept an almost continuous machine-gun fire across our trenches all the afternoon. Several times attempts were made to advance against us, but these were all easily frustrated. Our friend, Col. Spiller of the Divisional Artillery, was doing great work with direct shooting of his 18-pounder batteries on this afternoon. He hit the enemy very hard in the orchards where they were trying to collect, and no doubt in the edge of the wood also.

The break which had occurred further on the

right early in the morning was, however, the danger point, and it became very evident during the afternoon that the enemy was executing an encircling movement some two miles on our right. So little regard had the Hun at this stage for hiding his positions, that he had adopted the practice of making his forward troops fire Verey lights as they advanced to show his own gunners where they had reached. These lights at about 5 p.m. were going up well behind us on the right; the enemy was evidently approaching Le Quesnel from the south, and there was grave danger of a very large force being cut off. A hurried council of war was held, and the order to withdraw from the right of the Division was given. It seemed that it was none too soon. The whole Division had to get back through a bottle-neck, harassed all the way by machine-gun and artillery fire. The withdrawal was orderly, and by great good chance rain set in and made observation difficult for the enemy. Our own aeroplanes were also getting more in evidence.

Through the rain on this day we had a long and weary march to the Bois de Sencat, west of Castel, where the Brigade was ordered to concentrate. The last five miles of the march was across heavy country. The fatigue on that night after the strain of the day was almost the worst of the whole retreat. Scores of men were dead asleep as they stumbled along.

We arrived at the rendezvous at about 3 a.m., where we found our transport, had a hot meal, and lay down in the rain for the rest of the night. In the morning our Colonel was visited by the Divisional Commander, Maj. Gen. Daly, who thanked the whole Battalion for the work of the past eight days. This was the morning of March 29, and as there was no sound of battle very close to us, we felt we were in for a bit of a rest.

It appeared later that the enemy had taken Montdidier away south, and was thrusting on in front of us at a great pace. The Division was to go up in the evening to take up a position in front of Thezy from Berteaucourt to the Domart–Amiens road—the main road from Villers Brettoneux to Amiens. The 73rd Brigade were to be on the right, 72nd Brigade on the left, 17th Brigade in reserve. We moved up just before dark. The Entrenching Battalion took up positions in the line, which consisted of a series of disconnected posts, just dug by army engineers. We took up a position in the Copse behind them, where we proceeded to dig in. By 10 a.m. we had formed a series of posts just in front of the copse on a front of 1,200 yards, our left resting on the Amiens road. The enemy was reported to have advanced through Mezières, and was proceeding towards Moreuil and Castel.

There were now a considerable number of troops—French and British—between our positions and the enemy, and as there was no immediate threat to Thezy, the Battalion—less the East Surreys, who remained in the Copse manning a few of the posts—moved back to the village.

We got good billets in Thezy, and we were able to get a feed and some rest.

At dawn on March 31 we relieved the Entrenching Battalions in the forward posts. In the afternoon considerable activity commenced. The enemy had got into very difficult country now, and his lines of communication with his rapid advance must have been stretched near to bursting point, but he was making strenuous efforts for Amiens, the spires and towers of which he could now see in the distance.

The lines in front appeared to be wavering, but a series of counter-attacks was carried out between 4 and 7 p.m., and the position was more than re-established. The night was quiet and very cold, and we remained in position in the posts. Our artillery pounded the enemy's positions very severely all night.

At dawn on April 1 we were relieved by two companies of the Entrenching Battalion, and moved back to Thezy. The men rested most of the day and did all they could to clean up. There were a few civilians in the village and plenty of food. The day passed quietly and uneventfully.

On April 2 we were again at rest in Thezy. We were beginning to feel much fitter, and the past ten days felt like some horrible nightmare that is past. The Commanding Officer inspected all the companies during the morning. The day was quiet—hardly a shell had fallen in Thezy. Our own artillery cheerfully enough provided all the fireworks, and continued to shoot incessantly.

Clean underclothing and socks—a very welcome event—were issued to the men during the day.

At dawn on April 3 the East Surreys and the North Staffordshires relieved the two companies of the Entrenching Battalions in the forward posts. C Company were detailed for making an advanced Brigade Headquarters, and our indefatigable Staff-Captain and staunch friend, Capt. Ward, who had just returned that day from a course in England, organised baths, if you please, at the Château. We were getting entirely civilised.

The next morning, April 4, we did another side-slip northwards, to take up a position in support to the Australians behind Villers Bretteneux. At 4.30 a.m. we left Thezy and marched across the Amiens road through Gentelles to the Bois l'Abbé, where the Brigade concentrated. The Divisional Reinforcement Battalion, which had been with the 17th Brigade, relieved our forward posts in front of Thezy, and the concentration of the Brigade in the new positions was completed by 6.30 a.m.

The march through Gentelles was unpleasant. The enemy was just beginning his first big attack on Villers Brettoneux, and was shelling everything within reach which might be a billet. Thezy received a few just as we left, and as we marched through Gentelles several shells were about. One fell in the road as a platoon of A Company was passing, and we had eight casualties.

The forward part of the Bois l'Abbé received a lot of attention during the morning, and our old friend the gas shell was much in evidence. Fairly heavy rain set in, however, and the gas did little damage. The bombardment throughout the day was directed mainly against our batteries all along the front. The posts in front of Thezy —which were fairly in the midst of guns—had a bad time after we had left them.

In the course of the day it transpired that the enemy had attacked opposite Villers Brettoneux and Marcelcave and further south. He had made some progress, but the ground was very heavy, and in places he had been driven back. In the afternoon the 72nd Brigade, which was in support to the northern half of the 17th Brigade (Gentelles to Bois l'Abbé), moved back about 3,000 yards to a copse a mile or so east of Boves, where Brigade Headquarters was established. The weather had now properly broken, and the rain streamed down. C Company moved up during the night to a position in the Bois l'Abbé in touch with the 7th Londons, and the rest of the Battalion dug in in front of the copse. A number of cavalry officers and others dropped in to Battalion Headquarters during the night. There had evidently been some heavy fighting at Villers Brettoneux, but the result was good, and everyone seemed cheerful. They were all surprised to find us still in the line. Numbers of our visitors told us they had heard of the 24th Division on all sides, and had imagined that we were all killed long ago.

The next morning, April 5, broke chilly, but quiet. We were so soaked that it did not matter how much it rained. We had dug some very fair positions, and the cookers gave us quite a good breakfast.

During the morning we got the breath of a rumour that we were to be relieved. We paid not much attention to it, because we had heard the same thing at least three times before, and we were beginning to think we were in this business for the rest of our lives.

But this time it was true. Orders were issued from Brigade during the afternoon. We were to move off after dusk, and motor-lorries were to take us away back behind Amiens and then to goodness knew where. A significant thing, however, was that during the afternoon maps were sent up to us with Dieppe and St. Valerie and expanses of blue sea on them. Even then we were doubtful, and longed for 8 o'clock to come before the orders were cancelled. We wonder if the reader can imagine what our feelings were. On March 21 we were at the end of a long term—ten days in the line. Since then we had been fighting almost continuously—at most times in desperately tight corners. We were reduced in numbers to about 100 men and five officers. Our boots were worn through, our clothes were torn; mentally and physically we were as tired as men could well be. Up and down the line there were cases of men losing their reason during these fiendish days—and small wonder. So it was a very quiet, but yet a very triumphant little band that finally reached an old town miles west of Amiens in the small hours of April 6.

It had not taken many lorries to convey our Battalion home that night.

About midday we entrained and were taken down to St. Valerie. From there we had a ten-mile march—who minded that ?—to billets in the village of St. Blimont. The deathly stillness of the place was almost uncanny. We were splendidly received by the villagers. Although it was 2 o'clock in the morning, they tumbled out of bed to give us a welcome. Amiens had been saved—the Royal West Kents had arrived !

Many messages of congratulation were received by the 24th Division for their conduct during the March Retreat, notably from Lieut.-Gen. Sir H. E. Watts, commanding the Corps, from Lieut.-Gen. Sir C. T. McM. Kavanagh, commanding the Cavalry Corps, and from our two late Divisional Commanders, Maj.-Gen. Sir J. E. Capper and Maj.-Gen. Sir L. J. Bols.

The following honours were awarded to the Battalion for service between March 21 and April 5, 1918 :

BAR TO THE DISTINGUISHED SERVICE ORDER

Lieut.-Col. H. J. Wenyon, D.S.O.

THE MILITARY CROSS

Rev. G. D. Barlow-Poole, Lieut. G. S. Bowen, 2nd Lieut. J. B. Cryer, Lieut. E. O. Goulden, 2nd Lieut. B. Stephens, Capt. S. G. Thompson, 2nd Lieut. P. H. Tyler.

BAR TO THE DISTINGUISHED CONDUCT MEDAL

Regt.-Sergt.-Maj. R. Rankin.

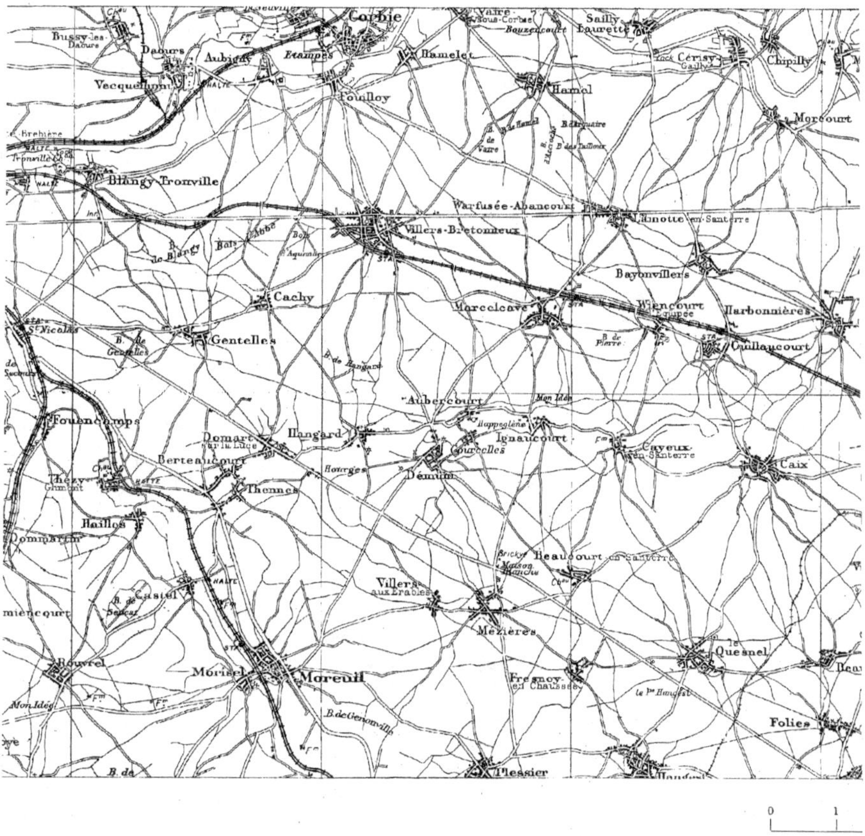
Corbie
Bussy-les-Daours
Daours
Aubigny
Vecquemont
Hamelet
Hamel
Fouilloy
Sailly-Laurette
Cérisy-Gailly
Chipilly
Morcourt
Blangy-Tronville
Warfusée-Abancourt
Lamotte-en-Santerre
Villers-Bretonneux
Bayonvillers
Cachy
Gentelles
Marcelcave
Wiencourt-l'Équipée
Harbonnières
Guillaucourt
Aubercourt
Ignaucourt
Fouencamps
Domart-sur-la-Luce
Hangard
Courcelles
Caix
Berteaucourt
Thennes
Démuin
Cayeux-en-Santerre
Hailles
Thézy-Glimont
Dommartin
Beaucourt-en-Santerre
Villers-aux-Érables
Castel
Mézières
Quesnel
Rouvrel
Morisel
Moreuil
Fresnoy-en-Chaussée
Folies
Plessier
B. de Genonville
0
1

Distinguished Conduct Medal

Cpl. E. H. Cherriman, M.M., Pte. A. E. Smith.

The Military Medal

L.-Cpl. G. Boylan, Pte. J. Bradley, Sergt. L. Green, L.-Cpl. F. Holford, Cpl. W. Jasper, Pte. W. Paine, Pte. J. Scripps, L.-Cpl. T. Shewry, Pte. R. Simmons, Sergt. C. A. Taylor, Pte. S. G. Wells, Sergt. C. Williams, Pte. J. Winder, Cpl. C. Woodrow.

The Meritorious Service Medal

Regt.-Quartermaster-Sergt. F. A. Ifould, Cpl. H. R. Verrall.

PHASE VIII

TRENCH WARFARE AND "PEACEFUL PENETRATION" AT LENS—DEAN'S V.C.

May—September 1918

ON April 10 Brig.-Gen. R. W. Morgan, D.S.O., inspected the Battalion on parade. He afterwards addressed the Battalion, and congratulated very warmly all who had taken part in the recent operations.

At St. Blimont we were reinforced by a draft of 222 and the return of 160 who had been on leave to England. The Battalion was finally made up to a strength of 850 rifles.

The work of reorganisation of companies was taken in hand immediately and completed by April 12. The remainder of the month was devoted to intensive training. The instruction of new section commanders and the training of a large number of new Lewis gunners was actively carried out. During the month our Lewis guns were increased to twenty-four, including four for anti-aircraft work.

The Battalion was fortunate in its drafts. About 40 per cent. were men who had had previous experience in the field; the remainder were men of the A4 class from England—about nineteen years of age and very keen.

On April 17 the Battalion moved to Ourton, near Pernes, where the Division was in General Headquarters reserve and ready to move at six hours' notice. Every third day the Canadian School Training Ground at Pernes was at our disposal, where there was an excellent 20-target rifle range, bayonet fighting courses, and drill grounds. At Ourton we made two rifle ranges, and much valuable musketry training was done.

A large draft of officers joined us at Ourton, including Capt. H. S. Brown from a six months' tour of duty at home. Capt. Brown became 2nd-in-command, and the Battalion organisation at this time was as follows :

Commanding Officer, Lieut.-Col. H. J. Wenyon, D.S.O.
2nd-in-Command, Capt. H. S. Brown.
Adjutant, Capt. R. P. Baker.
Transport Officer, Capt. J. N. Arnaud.
Quartermaster, Capt. J. R. Wood.
Company Commanders :
- A Company, Capt. J. H. Selfe.
- B Company, Capt. R. G. Rogers.
- C Company, Capt. S. G. Thompson.
- D Company, Capt. F. Proctor.

There was very little time for pure recreation now, but we got a holiday one Saturday, and Capt. Brown organised a very successful athletic sports meeting held on April 27. There was some very good running, one feature being the winning of the 100 yards officers' race by Gen. Morgan. The meeting did a lot of good, as the men were in danger of getting a little "stale" with the pressure of continuous training. The events in which mules participated caused considerable amusement.

A Brigade rifle meeting was held at Pernes on April 29. On the total points scored for shooting the Battalion was first, and on the points scored for field work, particularly section commanders' control, the Battalion was also very easily first. This was a very satisfactory feature, and was a direct result of very hard work by the officers and N.C.O.s during the previous weeks. Of the total prize money of about 550 francs, the Battalion took 385 francs.

On this day we heard rumours of a move to the line, the region of Lens being hinted at as our probable destination. On April 30 representatives were sent to reconnoitre the line in front of Lens—the very front, as it happened, on which the Battalion was fighting a year before when the enemy withdrew after the battle for Vimy Ridge.

On May 1 we found ourselves in old familiar spots—Sains-en-Gohelle, where we stopped for tea, Bully-Grenay, and Maroc, where we spent the night in cellars. At dusk on the 2nd we moved up the Fosse 11 road, along which the Battalion had fought its way in April 1917, into Cité St. Pierre, where we took over from the Canadian Cameron Highlanders, and became battalion in support. The 72nd Brigade took over the St. Emilie Sector. This had been a thickly populated mining district, and the trenches wound about through the ruins of old houses and under roads. The whole district was a mass of demolished houses, brick rubble, and debris of every description.

The front line ran just short of the north-western outskirts of Lens. Two main roads, the Lens–Bethune and the Lens–La Bassee roads cut through our trenches and joined in the city just behind the enemy's line. All our trenches here were old German lines captured a year before. There was a lot of the enemy's concrete work everywhere, chiefly in the shape of re-inforced cellars and houses and forward observation posts. One of these observation posts formed the inter-battalion boundary in the front line.

No Man's Land on the right was a tangled mass of ruins of every description: broken houses, old demolished mine buildings, wire entanglements, the whole churned up by shells in a remarkable way. On the left there were not so many buildings in front, and No Man's Land was old grass land, with here and there an isolated ruined house. On the left within our lines was the Canteen Tunnel, an elaborate underground work with about fourteen entrances, stretching from the support line, Douglas Trench, up to the front line, Chicory Trench, where an exit could be made into Mason's House, an old ruin just in front of our parapet. This tunnel was always more trouble than it was worth. All the entrances had to be guarded, and it became later the favourite hunting-ground of Hun raiding parties.

The fact that this had been once a flourishing and thickly populated district made it the most desolate-looking sector we were ever in, and yet one of the most interesting. It offered never-ending opportunities for exploration. Searching for good places for Lewis gun or rifle posts, cellars worth reinforcing as shelter for the men against bombardments, salving and building—all became sports which were developed to a high degree. Small wonder was it that this was the sector we shall always remember for its defence schemes, changing nearly every tour and baffling overwrought company commanders by their intricacies. At the end of our stay here, any company commander who could tell where every one of his sections was, and where they would be in case of attack, was established as a Pelmanist of no mean order.

The 72nd Brigade front at the outset was held by two battalions, with one battalion in support, but this was changed almost at once to one battalion in the line distributed in depth, one battalion in support at Cité St. Pierre, and one battalion in reserve at Bully-Grenay.

We started with three days at Cité St. Pierre. Battalion Headquarters was in a dug-out under a ruined house known as Hodson's House, and everyone in the Battalion had pretty good billets in the cellars of the ruined houses. During these three days we were reconnoitring the left battalion front which we were to take over.

On the 5th we moved up after dark and relieved the 1st Battalion North Staffordshire

CITÉ ST. PIERRE—THE OPERA HOUSE.

OLD GERMAN CONCRETE OBSERVATION POST IN OUR FRONT LINE OPPOSITE LENS.

In foreground—"No man's land."

Regiment. B and D Companies were in the front line, A Company in support, and C in reserve. Battalion Headquarters was well forward in a dug-out in Counter Alley, the main communication trench. The enemy was quiet, and we settled down at once to learn the new front. On the 8th we got alarming news from a prisoner captured on another front. Lens was to be the next jumping-off place for a big enemy offensive. He was to attack in force on the night of the 9th/10th on the front from Lens–Robecq. His bombardment was to commence at 10.30 p.m., and his attack, with twenty-two divisions, was to be launched at about 3 o'clock in the morning. May 8 and 9, therefore, we spent in fortifying our positions as much as possible. By the afternoon of the 9th all possible preparations were completed. Gen. Morgan came round our front during the afternoon. Every man was in his appointed place, and we settled down to a weary wait for the Hun to start. There was no " wind up " in the trenches. We felt we could do no more than we had done.

The great offensive did not take place. Why, we never found out, but we did learn that everything was in fact ready for a big attack on our front at this time. We waited patiently for 10.30 to come. After that we waited all night, and for several more nights, but nothing extraordinary happened at all.

On May 12 the Divisional front was rearranged, and was henceforth held by three brigades in the line.

The 1st North Staffordshire Regiment took over part of our line on the right and the 1st Royal Fusiliers (17th Brigade) relieved the rest of us.

On the night 13th/14th we moved out into reserve at Bully-Grenay. A Company were at Cité Calonne, about two kilometres away. The billets at Bully and Calonne were by no means clean or comfortable, and we commenced the work of cleaning up, building, and repairing in every direction.

On the 16th we moved up again to Cité St. Pierre in support. Two companies had battle positions in the red line, our main forward battle position here, and two companies were in brigade support to be used for counter-attack or reinforcement where necessary on the Brigade front.

During this tour at St. Pierre, Maj. Brown organised a raid on the enemy's trenches opposite Conductor Trench. The Lens front at this time was famous for gas, and at Cité St. Pierre we concentrated on gas-proofing the cellars and making them more comfortable. The Hun used mustard and sneezing gas on the forward and battery areas, and we indulged in retaliation in the way of projector attacks, gas-cloud attacks, and a certain amount of gas shelling. On the 22nd we moved up to the front line. The dispositions were now two companies forward in the outpost zone, and two companies back in the intermediate or Blue Line. On the night of May 24/25 A Company carried out the raid in conjunction with gas-projector attacks. This was the first occasion on which a raid had been attempted in conjunction with a projector attack. In view of the uncertainty as to the enemy's intentions here at this time, it was essential that prisoners should be taken, and we were very warmly congratulated on the capture of three very useful Huns in this raid. The following is the official account, sent in by Lieut.-Col. Wenyon at the time :

"The raid on the enemy trenches east of Conductor Sap took place last night.

"The whole party was assembled in the sap and in correct order by 10.50 p.m., and the end of the tape secured ready to move out.

"The party left the assembly position on commencement of the barrage at 10.54, and moved on a compass bearing of 153°. At 10.56 the gas-projector attack took place. A number of the cylinders fell in No Man's Land among the raiders, and, although they probably contained smoke only, they had a suffocating effect on the men, and the whole party got badly disorganised. 2nd Lieut. Treacher reached the enemy's wire with one or two men only. He found that it was possible to get through the enemy wire, but, as he was practically alone, he decided to come back.

"The party was back in our trenches at 11.13 p.m.

"I did not hear of the situation till 11.20, as the telephone wires forward of advanced Battalion Headquarters had been cut.

"At 11.25 Maj. Brown came to advanced

Headquarters and reported that the raiders had failed to enter the enemy trenches and were back in our line. The artillery barrage had been stopped.

"I decided that the raid should be done again with half the party at 2 a.m., when another gas-projector attack was due.

"A new barrage was arranged for 2 a.m., to remain on the enemy front line for one minute, then lift on to his support line. I saw the officers of the raiding party, and the Officer Commanding 6-inch Newton Battery. The men of the raiding party had been sent back to their posts in our trenches, and collecting them took a long time.

"Major Brown, with a runner, went out at about 1.40 from the assembly position and laid the tape to a point within thirty yards of the enemy wire. On his return he found 2nd Lieut. Wallis and 2nd Lieut. Piggott with one section ready in the assembly position.

"The projector attack which had been put forward ten minutes, and took place at 1.50, called forth some retaliation on our front line and hampered the assembly considerably.

"The party proceeded at once down the tape, and successfully got through the loose wire and entered the enemy's trenches. No opposition was encountered in entering the trench, and only one machine gun from the left was firing while the party was crossing No Man's Land—and that very wildly and without effect. On entering the trench, 2nd Lieut. Piggott worked to the left with four men, and 2nd Lieut. Wallis to the right with the remainder of the party. Maj. Brown and a runner took up a position at the point of entry as Raid Headquarters. As 2nd Lieut. Wallis started down the trench with his party, two of the enemy appeared round a traverse. They were at once fired on, and 2nd Lieut. Wallis threw a bomb into the next fire bay. The Huns gave a loud yell, and retired at the double, followed by 2nd Lieut. Wallis and his party. About forty yards down the trench 2nd Lieut. Wallis came to a dug-out entrance. He called down, but got no response; he then threw a Mills bomb down, and, following shrieks from the bottom of the steps, one man emerged. No signs of anyone else being inside could be found, so, after doing as much damage as possible to the dug-out and throwing more bombs down, 2nd Lieut. Wallis returned with the prisoner to the point of entry.

2nd Lieut. Piggott had meantime worked down about fifty yards to the left, and was having trouble at a dug-out with some fractious Boches. Hearing the noise, Maj. Brown sent his runner to investigate and assist. On arriving at the dug-out entrance, 2nd Lieut. Piggott had shouted down, and some guttural response had been given, but no inclination to come up. He then fired his revolver through the gas curtains, which were down, and the gas curtains were then raised. A Hun appeared on the steps, and was evidently inquiring what was the matter. 2nd Lieut. Piggott then threatened him with his revolver.

The Hun emitted a piercing shriek, and came forward with his hands up. Another Boche came to the bottom of the steps, and was extracted in a similar fashion. No more response could be got from this dug-out, although Mills bombs were thrown down. 2nd Lieut. Piggott then returned with his prisoners to the point of entry; 2nd Lieut. Wallis fired the green light—the signal for withdrawal. The whole party then returned to our line. They had been in the enemy trenches twelve minutes. No trouble from enemy machine guns or rifle fire was experienced while the party was crossing No Man's Land. Maj. Brown and 2nd Lieut. Wallis got back most of the tape from No Man's Land when they returned.

"The party which entered the enemy trench consisted of the following: Maj. H. S. Brown, 2nd Lieut. G. A. E. Wallis, 2nd Lieut. D. T. Piggott, L.-Cpl. Warford, Pte. Baker, Pte. Tilby, Pte. Smyth, Pte. Bonney, Pte. Harcin, Pte. Bowes."

The following decorations were awarded for the raid:

The Military Cross

Maj. H. S. Brown, 2nd Lieut. G. A. E. Wallis, 2nd Lieut. D. T. Piggott.

The Military Medal

L.-Cpl. H. G. Warford, Pte. J. Baker, Pte. E. Bowes.

Battalion Headquarters was now in the old Mine Offices near the Red Line, and just off the Bethune road, known as Cameron Castle. On the night 28th/29th we were relieved and returned to Bully-Grenay. On May 29 a terrible tragedy occurred. In the course of a football match the Hun dropped two bombs among the players, and we had terrible losses. Regt.-Sergt.-Maj. Rankin, Sergt.-Drummer Goldsbury, and Pte. Lloyd (Battalion runner) were men who in their own spheres it would be difficult to equal. Rankin had been in every fight. He had won the D.C.M. on the Somme, and two days after his death the news of the award of a bar to his D.C.M. for his matchless bravery on the March retreat was announced. It would be difficult to say how much the Battalion owed to the energy and devotion of its senior warrant officer. Sergt. Goldsbury had worked our drums up to a wonderful pitch of perfection. He was a ceaseless worker—his energy and cheerfulness were a byword in the Battalion. A splendid fellow in every way. Young Lloyd had been a runner for a long time. On the Somme, through the Ypres fighting of 1917, at Hargicourt, and right through the March retreat he had been one of the bravest of the brave at all times. Nothing was too hard or too dangerous for him to take on. He was always so cheerful and absolutely reliable. Two signallers were also killed. It was indeed a cruel fate that these men after facing death a thousand times should be killed at a football match far away from the fight. There were very few dry eyes at the Bully cemetery when we buried these brave men the next day.

June was on the whole a quiet month. Two deserters gave themselves up to C Company, and the enemy attempted two or three raids on our trenches unsuccessfully.

The month of June 1918 is chiefly remembered for the extraordinary epidemic of influenza which swept through the British Army—and it is believed in a worse form through the German Army. We lost temporarily during the month upwards of 300 officers and men through this scourge. For about ten days the Battalion was commanded by Maj. Esdaile of the Queens, when the Commanding Officer was on leave and 2nd-in-Command down with 'flu.

On the night of June 12/13 a party of B Company, under 2nd Lieut. Michie and 2nd Lieut. Binney, got into the enemy's trenches with the assistance of a Bangalore torpedo. The following is the report on the patrol :

" A fighting patrol of two officers and two sections entered the enemy trenches last night.

The tape-laying party (2nd Lieut. Michie and two other ranks) left our trench at 11 p.m. Proceeding beyond our wire by Conductor Sap, an enemy patrol of about sixteen was seen advancing on Mason's House. Our trenches were being trench mortared at the time. Our raiding party was temporarily got into position to repel this raid, and our two officers went forward again into No Man's Land to reconnoitre. On being approached, the enemy dispersed and made his

HODSON'S HOUSE.
Support Battalion Headquarters.

CAMERON CASTLE.
Front line Battalion Headquarters.

way hurriedly to his own line. The enemy was using gas in trench mortars. After this, 2nd Lieut. Michie went out and laid the tape successfully, from Conductor Sap to the enemy wire. A covering party (with Lewis guns) was then placed in position in No Man's Land, and the two sections and the Bangalore torpedo party advanced to the end of the tape on the enemy's wire. The torpedo was exploded successfully at 1.10 a.m., and 2nd Lieut. Binney led half the party through the gap into the enemy's trenches.

"Immediately the Bangalore torpedo exploded the enemy put up two green lights, and within a few seconds he opened up with trench mortars on No Man's Land and his own wire. Several trench mortars fell almost in the gap made by the torpedo and divided our patrol into two parts, inflicting two casualties on the rear section. 2nd Lieut. Binney, with the leading section, remained in the enemy trenches for some minutes and thoroughly investigated it.

"The trench still looks quite new, but is only about three feet deep, and there are no dug-outs in this vicinity. About 1.25 a.m. the order was given to return to our lines. This was done successfully without further casualties, although trench mortars were bursting freely in No Man's Land. From the speed and accuracy with which the enemy's barrage was brought down it would appear that the exact locality of the raid had been discovered. Probably the tape-laying and Bangalore parties had been spotted. 2nd Lieuts. Michie and Binney both did good work."

By the end of June, profiting from the experience in the recent German offensive, the principle of depth in the holding of the line was taken up thoroughly. This necessitated a further change in the defence scheme. By this time we were getting a little accustomed to the constant changes. In general terms the line was held now by one company only in the outpost zone, and three companies back in the main forward battle positions known as the Black Line, about 1,500 yards behind the front line. The object of this arrangement was that the large majority of the garrison should be beyond the range of the enemy's trench-mortar fire, which he had used in the opening stages of his offensive on so many fronts during the spring and early summer with annihilatory effect.

Battalion Headquarters was moved from Cameron Castle to some new dug-outs on high ground in Cité St. Pierre behind the Black Line.

On one memorable occasion, when we were in the front line, a practice "stand to" was ordered by Brigade. Gen. Morgan came up, and had gone round the Left Black Line company's front, when the enemy elected to put down a thick gas bombardment of the Black Line area. The inspection was abandoned, and the General and his party started to grope their way back in gas-masks. Each communication trench which was tried seemed to be a particular mark for the enemy's guns. Finally, by a very circuitous route, the party arrived at St. Pierre Church. Here all seemed quiet, but as soon as the General and the Brigade Major started to climb the hill to Morgan Mansion, Brigade Headquarters, gas shells began to follow them again, and continued in their tracks all the way home, the final insult being a perfectly good shell which landed at the entrance to their dug-out just as they were going to enter. In retrospect it looks to have been a very amusing evening. At the time it had its comic moments, but not many.

During July everyone was very busy. When we were out of the line we were training hard. Several days were spent at the training-ground at Marqueffles Farm, where there was a good rifle range, and where we had opportunities to practise with tanks and contact aeroplanes. When in support, large parties were out every day making concrete Lewis-gun and rifle posts in the Blue Line, where the support platoons of the outpost company had their battle positions.

In the front line we were patrolling with exceptional vigour, keeping touch with the enemy in every way possible. Several times we encountered the enemy in No Man's Land and drove him off, and he on his part made one or two attempts to get into our lines without success.

On July 20 the Corps Commander, Sir Aylmer Hunter-Weston, visited us and went round the Black Line. The enemy honoured the day with a very heavy bombardment of the area round Counter Alley from which our gas projectors were usually released. The inspection of the Black Line was a vigorous affair; few who took part in the race will ever forget it. The distinguished general kept up his habitual pace of six miles per hour without turning a hair, and by the end of the afternoon his staff and following were stretched over at least a mile of trench. Many of them had succumbed into one or other of our Company Headquarters for liquid refreshment very early in the contest. Of twenty starters in the Marathon, only four were in sight when the favourite flashed past the post at Battalion Headquarters at 5.30 p.m.

The reader will have taken for granted on reading the name of our Corps Commander, that in addition to the other strenuous work of these days, activity with the whitewash brush and other fly-proof devices was to be seen on all sides at all times.

On July 21 the first performance of the new Brigade Concert Party, " The Dragonflies," took place. It had been in the rehearsal stage for a long time, but the results fully justified the care which had been given to it. The Party continued to entertain us until long after the Armistice.

The Divisional artillery had a forward section of guns at the Opera House in St. Pierre at this time, and the enemy carried out several protracted shoots on the Opera House area. He never succeeded in hitting the guns, but parts of the village were very unpleasant at times. On the night of August 4/5, Battalion Headquarters was shelled heavily with high-explosive and sneezing gas, direct hits being obtained on the mess and Commanding Officer's dug-out. On

August 5 a particularly heavy shoot on the Opera House made a nasty mess of that part of the village. Lieut. Walker, an R.E. officer, who had done a lot of work with the Battalion, was killed. In the front line patrolling activity was very great, and the enemy was given no chance in No Man's Land. Only on one or two occasions was the enemy encountered, and he invariably made off immediately he saw any of our men.

On August 8 the officers beat the rest of the Battalion at cricket by 120-odd to 60-odd, Col. Wenyon being top scorer with some 60 runs to his credit. This was by way of revenge, as on the previous tour at Bully the other ranks had been successful by 110–102. On the 13th we played the Sherwood Foresters, and were rather badly beaten.

On August 9 the Divisional Horse Show took place. We took second prize in transport turn-out, and "Diana," the veteran C Company mount, won second prize in the officers' chargers competition.

A few weeks later in the Corps Commanders' Transport Cup Competition, when the entire transport of all units were inspected for efficiency and turn out, we were placed first in the Division.

Our Transport Section was justly proud of this result, and it was a well-merited reward for the splendid work of Capt. Arnaud, Sergt. Shepherd, and all ranks of the Transport Section.

It has been difficult in these notes to mention the work of our Transport from month to month, but this has not been because it did not deserve mention. Everyone intimately acquainted with the Battalion would agree that the work of the Transport Section was always one of the great features of our life. Capt. Arnaud joined the Battalion on the Somme in 1916, and remained Transport Officer till the end of the war. He was associated first with Sergt. Wood, M.S.M., and later with Sergt. Shepherd, M.S.M. The whole section pulled together magnificently, and their loyalty to the Battalion was always most marked. During the very difficult days in the Salient in 1917, throughout the winter at Hargicourt in 1917 and 1918, and at Lens during the summer of 1918, the men in the line had ample reason for thankfulness to Capt. Arnaud and his splendid section. But the greatest work was during the retreat in March and April 1918, and during the final advance from Cambrai in October and November 1918. During both these times of trial everyone was fed and supplied every day; and during the March retreat especially, the fight the Battalion was able to put up day after day was only made possible by the indefatigable zeal of our Transport Section so gallantly led by Capt. Arnaud. By hook or by crook he succeeded every day in getting hot meals up to the men who were fighting, and supplies of ammunition were rushed up at a moment's notice wherever they were needed.

The Military Cross awarded to Capt. Arnaud and the M.S.M.s awarded to Sergt. Shepherd and Cpl. Verrall were among the very best earned rewards gained during the whole history of the Battalion.

During August the Hun commenced to shell Fosse 10, where up to that time the Transport and Quartermaster Stores had spent a very peaceful summer. It had previously been subject to night bombing raids, and a few long-range shells every now and then, but at the end of August he was shelling there daily and making things very unpleasant. On the 23rd he got a direct hit on the Quartermaster's Mess, and on the 24th the Transport moved, followed on the 28th by the Quartermaster's Stores, to Aix-Noulette. Unfortunately the Hun then turned his attention to Aix-Noulette, and it became nearly as uncomfortable as Fosse 10.

Some time before this, during one of the bombing raids on Fosse 10, we lost Regt.-Quartermaster-Sergt. Ifould, who was most unluckily hit by a splinter, which came through his window. He had succeeded Regt.-Quartermaster-Sergt. Dadd in the previous summer, after being Coy.-Quartermaster-Sergt. of B Company for a long period. He was very popular, and was a most valuable warrant officer, and we were all very sorry to hear of his departure to England. Coy.-Sergt.-Maj. W. Dunk, D.C.M., was his successor.

Bully-Grenay was also getting quite uncomfortable, and was frequently either bombed or shelled, mostly at night.

On the 27th, when we were at Bully, we suddenly had orders to go into the line to the right of our Brigade front for one night. The front was being rearranged, and the 73rd Brigade was moving round to take over from the left brigade of the 20th Division in front of Lievin. The enemy shelled Lievin heavily at night with mustard gas, and the reserve company had a few casualties. We returned to Bully the following night.

On the 29th a special performance of the Brigade Concert Party was held in aid of our Regimental Prisoners of War Fund, the proceeds of which, 903 francs, were duly sent to the Depot.

On the 30th a most successful Sergeants' Mess Concert was held, with Regt.-Sergt.-Maj. Greenaway in the chair, to celebrate the close of the Battalion's 3rd year in France.

Towards the end of August there was a distinct feeling that if anyone was going to do a push at Lens, it would not be the Boche. On September 1 the system of "peaceful" penetration commenced. 2nd Lieut. R. N. Killick, with a platoon of C Company, pushed forward at about 3 o'clock in the afternoon into the ruined streets of Lens. This small party were the first British troops to enter Lens, and once there they established positions for D Company, who relieved C Company at night, to take up and hold.

We had a busy time from now onwards. Although our line had been advanced only a few hundred yards, it was decided that working parties could show themselves in hitherto forbidden places by day. With a view to a possible advance, all work was concentrated on the clearing of the two main roads in Lens, the making of forward tracks, and the formation of large forward dumps of ammunition and rations.

The enemy strangely enough left these huge working parties almost unmolested, and in a very few days we had the strange sight of almost perfect *pavé* roads running practically into the front line, and men employed in the open in all directions in broad daylight.

In the front posts which we had taken, things were not so quiet. The enemy would put down heavy bombardments on to place after place each day. On September 3 he attacked the East Surrey posts at the junction of Claud and Canary Trenches on our left, captured two days before, and recaptured them. In rendering assistance in this fight, one of our officers, 2nd Lieut. F. Carville, M.C., was badly wounded. We remained in the front line until September 12, during which time inter-company reliefs took place each forty-eight hours. A great deal of work had been done, and we handed over a new organised front line to the 9th East Surrey Regiment.

Our six days in reserve at Bully were busy ones. On the 15th the whole Battalion attended a tank demonstration at Marqueffles. On the 16th we carried out a battalion tactical exercise in the form of an advanced guard march to Marqueffles and an attack in open country. Field guns, machine guns, and Stokes mortars co-operated.

On the 18th we moved into the line again in relief of the 9th East Surrey Regiment. The map references of the forward posts taken over did not agree with the actual posts held, and after a few quiet days and some nights of active patrolling it was decided to put this right. On the night of the 23rd/24th, a party of C Company retook the enemy's post at the junction of Canary and Claud Trenches, which post had been disputed several times during the past weeks. Up to now it had never been held for more than a few hours, and it was now our turn to have a shot at taking and keeping it. The full account of the events of the night, and the first attempt of the enemy to recapture the post on the morning of the 24th, are as follows :

When the position at the junction of Canary and Claud had been captured, consolidation was carried on vigorously all night, and by 6.30 a.m. considerable improvement had been made in the positions. At about 6.40 a.m. the enemy started to shell our old front line and communication trenches heavily with 5·9s, and shortly afterwards the officers in this area, 2nd Lieut. Davis and 2nd Lieut. Beynes, who were at the time engaged in establishing a fresh post to connect up Mason's House with our new position, heard heavy bombing from the direction of the latter. A party of the enemy, about forty strong, had entered the post from the north-east, one lot coming straight down Canary and another entering Claud at its junction with Twisted Alley. The post was occupied, and one N.C.O. and five men were captured. 2nd Lieuts. Trenchard-Davis and Beynes at once organised a counter-raid. A bombardment with rifle grenades was commenced from a position in Canary, and the two officers each led a party along either side of the trench, ejected the enemy, and reoccupied the post. The Huns retreated hurriedly up Canary and over the open, and were followed with Lewis-gun fire from the recaptured post and Counter Sap. One German was killed just beyond the post, and it seems probable that several other casualties were inflicted on the enemy in his disorderly retirement. L.-Cpl. Binks and one of the men who had been captured were left, with a wounded Hun, in the charge of another Hun. When the attention of the latter was attracted by the commencement of our counter-raid, L.-Cpl. Binks relieved him of his rifle and brought in his escort and the wounded Hun as his prisoners.

The position was completely restored by 7.30 a.m. Our casualties were four men missing and two slightly wounded, and the enemy is known to have lost one killed and two prisoners and probably lost many more.

2nd Lieut. Trenchard-Davis and 2nd Lieut. Beynes acted with great resource and courage in organising and carrying out the counter-raid, and L.-Cpl. Binks showed great presence of mind and determination in turning the tables on his captors.

On the evening of September 24 a platoon of D Company relieved the C Company garrison in Canary Trench. The D Company platoon, No. 16, was under the command of Lieut. D. J.

Dean. Most of the men in the platoon were young soldiers, as yet almost untried in the real rowdy-dow of a fight. As they took up their posts in that battered trench, none of them knew that in the next two days they were going to write the most glorious page in the history of their Battalion. They were heavily attacked almost immediately they had taken over the post, and for two days and two nights after this they were constantly attacked by overwhelming numbers, and with all the fury of shelling and trench mortaring which a resolute and exasperated enemy could summon to his command.

When No. 16 Platoon took over the post at 9 p.m. on September 24, it still required a great deal of consolidation, and the trench was very shallow. Within ten minutes of the arrival of the new platoon, the enemy made his second attempt that day to recapture the post. Covered mainly by machine-gun fire, the enemy advanced over the open from the north-east, but were driven off by Lewis-gun and rifle fire from the post. Lieut. Dean was in the most exposed positions, and moved about in the heavy machine-gun fire with absolute fearlessness. Immediately this attack had been dealt with, feverish work was put in to deepen the trench and improve the blocks in Canary and Claud. The enemy kept up a heavy machine-gun fire on the post, from which the trench, in most places only two feet deep, offered very little cover.

Just after midnight another attempt was made by the enemy. Two men were wounded,

CONCRETE PILL BOX.
Our First Platoon Headquarters in Lens.

DEAN'S POST.

and the enemy got so close to the post that it was in great danger. Lieut. Dean, however, led his men forward, and the enemy, surprised by this bold move, were driven off.

Again it was a case of digging for their lives, and by the time the dawn arrived both officer and men were nearly exhausted, but they were not to be allowed to rest. At 6 a.m. they were subjected to a third and very resolute attack by the enemy. A barrage of heavy shells was put down behind the post to cut off all communication, and the post itself was severely trench mortared. Lieut. Dean at once got his men into position, harassed the enemy approach down Canary with rifle grenades, and brought withering rifle fire to bear on them as they advanced over the open. One of the enemy was killed as he was entering the post down Canary Trench, and several casualties were inflicted on them as they retired in disorder. Regardless of danger, Lieut. Dean followed the retreating enemy over the open, but was driven back by rifle and machine-gun fire before he could secure any prisoners. For hours after this, enemy parties could be seen moving about their trenches. At first it was thought that they were coming up for a fresh attack, but there appears little doubt that actually they were engaged in removing casualties.

Throughout the 25th and the night 25th/26th Lieut. Dean and his men gave themselves no rest, but pressed on with the work of consolidation in face of great enemy opposition, and by dawn on the 26th the position had been very greatly strengthened. The garrison of the post was then relieved by two fresh sections of the same platoon, and Lieut. Dean remained in command.

At 10.15 a.m. on the 26th the enemy commenced to shell the post and its vicinity with 5·9-inch howitzers, and also put a very heavy barrage on all communications behind. Dean at once got his men into their fire positions, and a fresh attack by the enemy was beaten off with loss. When this attack was over, Dean went to visit his other posts, and during this time the enemy recommenced shelling with the greatest intensity, and obtained several direct hits on the post itself. In view of the casualties, the sergeant in charge decided to withdraw temporarily, and got the men back to a position some fifty yards from the actual post. The enemy immediately followed up his barrage and occupied the post, showing great resolution. Lieut. Dean grasped the situation at once. He collected his two sections, and led them with the utmost gallantry across the open, and the whole thing was done with such speed and determination that the enemy did not wait for the bayonet, but evacuated the post in the greatest disorder, suffering many casualties from our fire. This counter-attack was greatly assisted by the platoon on the right in Cinnabar Trench. Immediately 2nd Lieut. Cambrook saw Lieut. Dean preparing to counter-attack, he collected two of his own sections and led them under heavy fire over the open with a view to taking the enemy in the flank and cutting off his retreat.

He had reached a point on the flank of the enemy post before he was severely wounded in the chest, but his work was then done, as the enemy, seeing himself attacked from two sides, had already vacated the post. Lieut. Cambrook's quick decision and fearless leadership were of the greatest value, and were rewarded with the Military Cross, but unfortunately he died of his wounds a few days later, and never heard of his honour. His sergeant, Sergt. Alderman, remained with him until the post was captured, and then most gallantly carried him back over the open, and was rewarded with the Military Medal.

This was the last of the enemy's six attempts to retake the post in just over forty-eight hours, and five of the attacks had been withstood by Lieut. Dean and No. 16 Platoon of D Company. During the next night two dead Boches were brought in from the vicinity of the post, and groanings during the day testified to other enemy casualties. In addition, scattered round the post sufficient rifles and equipment were found to provide for a whole platoon, giving further evidence of extensive enemy casualties and of the completeness of his demoralisation. It should be mentioned that during the first bombardment on the morning of the 26th the enemy hit and exploded two heavy mobile charges, which had been brought up for the raid to be carried out in the afternoon. This explosion severely shook Lieut. Dean, who was very close at the time, and his subsequent work in organising and carrying out the counter-attack is all the more remarkable.

To sum up the work of Lieut. Dean, we give the closing paragraphs of his official recommendation for the Victoria Cross :

" During these two days Lieut. Dean worked unceasingly and with utter contempt for danger in an exceedingly difficult situation. His garrison was suffering casualties continually, but he himself bore a charmed life.

" Five times in this period he was attacked (three times very heavily) by a resolute enemy, and each time the attacks were thrown back.

" Throughout he inspired all his men with his own contempt for danger, and in consequence they all fought with the greatest bravery, though most of them were very young soldiers with little experience.

" He set an example of personal gallantry, initiative, splendid leadership, and devotion to duty of the very highest order."

The battered remnants of No. 16 Platoon came back on the night of September 26. Even then they were reluctant to leave the post in other hands.

The immortal defence of Dean's Post was talked of from one end of England to the other when the announcement was made that the King had awarded the Victoria Cross to Lieut. Donald John Dean. The Battalion heard the news of the award in December in a little Belgian village, after all fighting was over. The news was telephoned through to us when the whole

Photo by] [*Wykeham Studios Ltd., London.*

LIEUTENANT D. J. DEAN, V.C.

Battalion was in the village school at a Battalion concert. Such a scene of tumult and enthusiasm as greeted the announcement is impossible to describe. The events of those September days were still fresh in our minds, and a few of Dean's platoon were still with us. Every man in the Battalion during those frenzied moments of cheering took Dean to his heart, and felt that he was the man who would stand at the head of the Battalion for all time. Dean himself was, of course, not there. He was only just past the critical stage following the severe wounds he sustained during October.

The official recognition of the work of No. 16 Platoon was expressed by the awards of the D.C.M. to Sergt. Skipper, and the M.M. to Cpl. Eversfield, L.-Cpl. Goodwin, and Pte. Yates.

After the second day of Dean's defence it was obviously desirable to push the enemy away from the posts which always threatened ours. A daylight raid on the enemy post at the junction of Cinnabar and Claud was, therefore, planned for the afternoon of September 26.

The raid was led by 2nd Lieut. R. N. Killick, D.C.M., and 2nd Lieut. C. Trenchard-Davis with the utmost gallantry.

The following official account of the raid was written at the time :

" The raid on the enemy trench at the junction of Cinnabar and Claud took place this afternoon. The right party entered the enemy trench at the junction with Claud and the left party went over the open towards Cuthbert. 2nd Lieut. Davis, leading the left party, saw one of the enemy manipulating a machine gun just north of the junction of Cinnabar and Cuthbert ; he appeared to be putting in a new belt. 2nd Lieut. Killick, seeing the danger the left party were in, rushed the post and shot the German dead just as he was starting to fire. Immediately afterwards 2nd Lieut. Killick was hit in the head and leg by a light trench mortar or rifle grenade.

"2nd Lieut. Davis led his party 50 to 60 yards up Cuthbert. No enemy and no dug-outs were seen. He then returned, and proceeded up Cinnabar about 100 yards. At the point where the machine gunner was killed he found a two-entrance deep dug-out.

"Several steel helmets and overcoats were hanging at the entrance. 2nd Lieut. Davis shouted down, but there was no response. He then threw in two Mills bombs, after which there appeared to be a disturbance downstairs. He shouted again to the men to come up, but there was no response. The R.E.'s and carrying party with the mobile charges were then called up and the charges successfully exploded at the bottom of the stairways. 2nd Lieut. Davis inspected both entrances before leaving. The whole dug-out appeared to have collapsed, and both entrances were effectually sealed. There is no doubt that there were inmates in the dug-out, and that they were irretrievably buried there.

"As his task was finished, 2nd Lieut. Davis ordered the party to return. They brought back the body (which was that of an officer) and the machine gun (light) complete.

"Our casualties were 2nd Lieut. Killick and one other rank wounded. Before the raid commenced, two men of the trench garrison were killed and three wounded by enemy trench mortars."

Both 2nd Lieut. Killick and 2nd Lieut. Davis were awarded the Military Cross.

The following special Battalion Order was issued :

8th Service Battalion Royal West Kent Regiment

In-the-Field, September 27, 1918

SPECIAL BATTALION ORDER

During the past few days the Battalion has passed through experiences which have tested the soldier-like qualities of all ranks to the utmost degree.

You have stood the test in the finest possible way, and the fact that our line still remains intact and stronger than ever, and that very heavy casualties have been inflicted on a brave and resolute enemy, redounds very greatly to the credit of all concerned.

I wish to mention the magnificent defence put up by D Company during two days of very great stress and difficulty.

Five times a determined enemy, supported

by all means at his disposal, has attempted to drive in our post, and five times he has been thrown back with heavy losses.

I also wish to mention the very excellent work done by all ranks of C Company: first in capturing the positions from the enemy which we now hold; secondly, for re-establishing our post when a strong attack by the enemy had temporarily driven it in; and lastly, for the very gallant manner in which the successful raid was carried out by them in co-operation with D Company on the afternoon of September 26.

All the companies have done a tremendous amount of work in consolidating the new positions and in daring patrol work.

A and B Companies on the right have been subjected to the trial of heavy shelling in their posts on several occasions. I know that the Officers Commanding and all ranks of A and B Companies will be the first to join me in congratulating D and C Companies for their magnificent work.

With the certain knowledge that soldier-like qualities of the highest order exist throughout all the companies, we can look forward confidently to the future. Whatever tasks lie in front of you will be carried out to the utmost of your powers.

The fine work of the past few days is worthy of the very best traditions of our magnificent regiment, and the Queen's Own will always be proud of the officers, N.C.O.s, and men, many of whom have laid down their lives, who have done such sterling work during this week of stress.

H. J. Wenyon, *Lieut.-Col.*,
Commanding 8th Service Battalion
Royal West Kent Regiment.

Battalion Headquarters,
September 27, 1918.

On the right company front during these days the enemy artillery fire had been very heavy, particularly in the vicinity of a post known as the Zeppelin. This was in the ruins of some dilapidated factory, the twisted iron of which stood up very much in the shape of the framework of a large airship.

On the night of the 27th/28th, after shelling this neighbourhood heavily for twenty minutes with 5·9's, the enemy attempted to raid the post on the right of the Zeppelin. The post was held by B Company, and when the shelling started, about midnight, 2nd Lieut. C. P. H. Manley, who was the platoon commander, got his men into a dug-out for shelter. Immediately the artillery fire lifted, 2nd Lieut. Manley got the men into position, and seeing a party of the enemy at very close quarters, he got the Lewis gun into action himself, while the men used their rifles. Immediately after this, bombs and rifle fire came from the north, and 2nd Lieut. Manley saw another party of the enemy approaching him from that flank. He at once moved his post a few yards further south to a point from which he could command both raiding parties. From this point he brought rifle and Lewis-gun fire to bear on the enemy, who, after throwing one sulphur bomb into the now unoccupied dug-out, scattered and withdrew in considerable disorder. Just at this minute 2nd Lieut. Manley was badly wounded in the neck, but the enemy, who were in much greater strength than our post, had already been defeated. On searching the ground afterwards, one German rifle and five mobile charges were discovered near the post, and it is believed casualties were inflicted on the enemy, but the night was very dark and nothing definite could be seen. The men supported 2nd Lieut. Manley splendidly, and the next post also joined in with rifle fire during the enemy's retirement.

2nd Lieut. Manley handled the position with great resource, and by his quick grasp of the situation and fearless leadership undoubtedly frustrated a very dangerous attempt to capture his post. He was awarded the Military Cross, but never heard of his honour, as he died of wounds six days after the raid. Very quiet in his manner, he was a most excellent officer, and was very popular. His death was a great loss to his company, and to the Battalion.

It was at about this time that we lost the services of our Adjutant, Capt. R. P. Baker, M.C. He seems to have developed some form of blood poisoning, and had to go into hospital. We all expected him back in a week or two, but it was more serious than we supposed, and he never rejoined the Battalion. He came out as a Platoon Commander after the Battle of the

Somme, assumed the duties of adjutant in November 1916, and held the appointment for over eighteen months. A general officer once called him "a b—— half-baked civilian," because their views on some point did not quite agree, and Baker declined to alter his to suit the general's but it was in reality a very erroneous estimation of his military qualities. He was a good disciplinarian and a hard worker, and, unlike some adjutants, did not confine his activities when in the line to an office in a safe dug-out. His chief hobby was salvage. He was awarded the Military Cross in 1918 for consistently able and gallant work over a long period. 2nd Lieut. C. F. C. Mackaskie, who had been acting as assistant-adjutant, took over the duties of adjutant.

We had just previously said good-bye to our Padre, who was promoted to be senior chaplain to the Naval Division. Rev. G. D. Barlow-Poole joined the Battalion at Blackdown Camp just before we sailed for France, and he remained with us for three years. He was not exclusively our own property, as he was also padre to the East Surreys, but he always took a very keen interest in all phases of the life of our Battalion, and his departure was keenly felt by all ranks. He was indefatigable in his work for the comfort and welfare of the troops, and was very popular, which is perhaps the criterion by which a padre's work can be best judged. He was frequently to be found wandering about the front line, and performed most gallant service throughout the

March Retreat, for which he was awarded the Military Cross.

From the 27th onwards the enemy commenced to blow up his dug-outs. He had finally given up all idea of ejecting us from our advanced positions, and he could see that we intended to advance still further. Such a clear instance of the enemy being defeated utterly in a straight fight is difficult to imagine, and it was all done by the matchless gallantry of a mere handful of men. Our whole Division was on the point of being relieved, but before we left Lens for the last time at the end of September, it was obvious that the enemy was preparing rapidly for a retirement all along this front. His shelling at close range ceased entirely, and the destruction of his defences on all sides went on continually.

On the night of September 29/30 we were relieved by the 2/24th London Regiment, who had just arrived from Palestine.

It was a dog-tired but triumphant Battalion that marched down the old Fosse 11 road to Bully-Grenay that night.

We expected, correctly, that we should soon be in the fight again, assisting in the hunt of a retreating enemy further south. All this would be new to us, yet we looked forward with confidence to anything that might come along. The Battalion had been tried in the fire for a full month, and in a perfect inferno for a week. It had come out tempered and steeled.

From Bully-Grenay we went in buses to

Coupigny. The next morning we entrained at Hersin, and arrived that night at Sus-St.-Leger, a small village near Doullens, for a few days' rest and preparation for the final phase of the war.

PHASE IX

THE FINAL ADVANCE FROM CAMBRAI TO MAUBEUGE

October—November 1918

AFTER five days' training at Sus-St.-Leger—chiefly on the lines of open fighting in co-operation with artillery and machine guns—we entrained on Sunday, October 6, detraining at Ribecourt from the first troop train to run into the station. We marched to the ruins of Graincourt and bivouacked in fields for the night, among constant visits from enemy aeroplane bombs. On the following day positions were reconnoitred at Rumilly, and we moved up into trenches at Cantaing for the night.

At this time the British positions near Cambrai were, roughly, the same as those reached in the advance of November 1917.

On Tuesday, October 8, a general attack on the enemy's Cambrai positions in conjunction with other attacks on a wide front opened at dawn. The 72nd Infantry Brigade was in close support to the 63rd Division, and we were assembled in trenches near Rumilly at 8 a.m.

The attack immediately in front of us on the village of Niergnies was successful. In spite of several counter-attacks during the day, the village was still in the hands of the 63rd Division when the 72nd Brigade moved up to relieve them at night. The 1st North Staffordshire Regiment and the 9th East Surrey Regiment took over the line, and we were in close support behind the village. The enemy shelled the area just behind the village very heavily with gas shells all night.

About 9 p.m. we received orders that the Battalion would attack the village of Awoingt at dawn the following morning. The orders were received over the telephone, and details were very meagre. Battalion Headquarters had been pushed forward to a new position, where there was a telephone, close to the village. The intense blackness of the night and the heavy gas shelling made it most difficult to get in touch with the companies, and the final arrangements for the attack were not completed until nearly 4 a.m., when the written orders reached us.

The operation order was rather historic. It opened as follows : "The 8th Royal West Kent Regiment will capture the village of Awoingt at 5.20 a.m. on October 9, 1918. The Guards' Division will attack on their right, the Canadian Corps from north of Cambrai on their left."

The object of the attack was the final cutting off of the town of Cambrai, which lay away on our left burning all through the night and the next day. The assembly for our attack was successfully completed by 5 a.m. and our barrage opened at 5.20. The barrage was very heavy, and, apart from one or two isolated guns which were shooting short, very accurate. B Company attacked on the right in touch with the Guards, C Company on the left with D Company in support, and A Company in reserve. The enemy was not holding the village in great strength, and all objectives were taken without difficulty before 8 a.m. The captures by the Battalion included thirty-five prisoners, two field guns, three tanks, one limber with horses complete, a number of machine guns, and several dumps of ammunition. We had a few casualties from our own shells falling short, and a number from the enemy's heavy shelling of the railway after the objectives had been taken. A strong enemy detachment with machine guns had been left behind at the railway bridge just south of the village, and they were making desperate efforts to destroy it. A party from D Company, under Lieut. Dean, captured the whole detachment with machine guns, and so saved the bridge intact.

The attack, though it did not involve heavy fighting, was a splendid exercise for the Battalion in following a hastily arranged barrage and in keeping direction in open country. The leading companies were on the heels of the barrage all the way, touch was maintained with the Guards throughout, and direction was never lost. It was almost dark when we commenced to advance, we could not see the village, and we went forward on a compass bearing. Soon, however, the sun came out and the morning mist cleared, and we wish we could have had photographs of the

AWOINGT.

HAUSSY.

The house at the corner was Battalion Headquarters. The ground attacked over was beyond the village.

plain at about 7 o'clock that morning. The whole Battalion was spread out in extended order over our front of 1,000 yards. In front of the leading companies was the steadily advancing line of the creeping barrage, in front again, round the near outskirts of the village, smoke shells were bursting in profusion, while our big crumps falling continually in the village were sending up huge clouds of red dust from the unfortunate houses. On our left flank the town of Cambrai seemed to be one huge bonfire. It was the most wonderful view of a battle we ever saw during the war. Our casualties were 2nd Lieuts. Winch, H. A. Green, and nine other ranks killed, and 2nd Lieut. Taylor and forty-eight other ranks wounded. We remained on the objectives at Awoingt.

The village had been considerably knocked about by the shelling, and the enemy continued to add to the damage by his counter-shelling during the rest of the day. The 73rd Brigade passed through us in the afternoon and took over the advanced guard work. On the following morning the 17th Brigade followed the 73rd, and we became brigade in Divisional Reserve. During the day, October 10, we moved from Awoingt to Cagnoncles, where we were very comfortably billeted for the night.

The advance continued on the 11th, and we moved forward in the early morning to positions south of Rieux, where we dug in in artillery formation and remained for the night. It was during this digging in that Lieut. Dean was wounded. He was loudly cheered by the Battalion, and particularly by No. 16 Platoon, as he was carried away on a stretcher. This ended Dean's active service with the Battalion, as after a critical time in hospital in England he did not become convalescent until long after hostilities had ceased, and he was not sent out again. After the charmed existence he had had during the past weeks, it was the irony of fate which decided that he should be the only casualty in the Battalion on the day when he was wounded.

On October 12 the enemy was reported to have retired across the River Selle, and the 17th Brigade were following. The enemy had, however, left strong rearguards behind, and the 17th Brigade had very heavy casualties during the advance on St. Aubert.

At night we moved forward into billets in Avesnes.

In the early morning of the 13th we relieved the Rifle Brigade on the Selle river in front of Haussy. The 1st North Staffordshire Regiment were holding similar positions on our left, and the 9th East Surrey Regiment were in reserve in St. Aubert. Our forward positions were on high ground just short of the ravine through which the Selle flowed. The enemy shelled us persistently in these positions, and the North Staffordshire Regiment had heavy casualties from gas in Montrecourt Wood.

During the night of the 13th/14th our patrols were very active along the line of the river. They found all bridges had been demolished, and the enemy was holding the line of the railway beyond with machine guns in considerable strength. During the early morning of the 14th small patrols from A Company succeeded in penetrating into the village of Haussy. The village was strongly held by the enemy, and the civilians were still there. Our patrols were immediately reinforced by two platoons, the men crawling two at a time behind a low hedge into the village. Capt. Selfe and a few men succeeded in crossing the river at one of the broken bridges, and brought back three prisoners as a result of some hand-to-hand fighting in the streets. The civilians were very much on our side. They gave us information as to the whereabouts of the enemy, they concealed our men in their houses, and fed them on coffee and eggs.

In the afternoon the enemy realised what we had done, and attacked our new positions in the village, but we succeeded in holding on. Lieut. Piggott and Sergt. Ashdown were killed.

At a brigade conference held later in the day, plans for an attack on the river line were discussed. The attack was to be carried out by ourselves and the North Staffordshire Regiment, and was to have as its objective the high ground from the chapel in the northern outskirts of the village northwards to the railway opposite Montrecourt. The village itself was not to be attacked. Our attack on the right was complicated by several sunken roads and a sand pit, and we decided to try a new experiment in the nature of the artillery barrage to be used, in order to avoid a frontal attack on the sunken roads. The attack was to take place in the early morning of October 16.

We spent the 15th in the same positions. The ground leading to the assembly positions was thoroughly reconnoitred by officers and N.C.O.s, and all arrangements for the attack were completed by midnight. 2nd Lieut. Thomas, who spoke French like a native, went into the village at night, and 150 civilians were escorted back behind our lines.

It was decided to make an effort to assemble for the attack on the far side of the river. To do this it was necessary to ensure that there were no enemy outposts remaining between the river and the railway. During the early part of the night of the 15th/16th, therefore, harassing fire by our artillery was brought down on the railway and on the ground between the railway and the river. From 3 a.m. on the 16th this fire continued on the railway only, and a strong patrol from C Company (under 2nd Lieut. Wallis) advanced across the river and cleared the ground as far as the railway, and took up a line as close to it as the shelling would permit. There was only one temporary bridge available, and all our assembling troops had to cross at this point. From about 3.30 a.m. the enemy commenced a counter-preparation on a wide front. All roads and approaches to the village and river were shelled with gas. In spite of this, and largely on account of the thoroughness with which the reconnoitring had been done, and thanks also to the work of 2nd Lieut. Wallis and his covering party, the two companies, D on the right and B on the left, were successfully assembled across the river by 5 a.m. Owing to the lack of bridges, and the winding nature of the river, the assembly positions were on the left flank of our attack, but the special barrages which had been arranged assisted us in launching the attack from these positions.

The barrage opened at 5.10, and our men advanced to the railway without difficulty. It was then discovered that the battalion who were to have attacked on our left had failed to assemble on account of the enemy's shelling, and the left flank of our attack was therefore in the air. The enemy had several machine guns mounted on the railway north of us, and B Company suffered heavy casualties in crossing the railway. A large number of prisoners were captured in the shelters behind the railway embankment and in the sunken roads and the sand pit.

The enemy was obviously surprised by the novelty of the barrage, combined with an attack from the flank. In spite of the opposition from the left flank the attack progressed, and by 7 a.m. all objectives had been taken, together with over 200 prisoners, many machine guns, and a few trench mortars.

Our successful penetration into the village two days before, and our capture of one bridge-head, had led the Corps to include the village itself in the objectives of the attack, and the 9th East Surrey Regiment were detailed for this work. The greater part of the village was successfully taken, and we were in touch with their posts at the chapel.

We had a very long exposed northern flank, and it was necessary to put the greater part of our strength to protect it. A protective barrage was arranged to assist in this, and by 10 a.m. a defensive position had been made which seemed satisfactory. Had we had the men, and had the attack on our left been carried out, we could have exploited our success across the open country east of us, and could probably have got among the enemy's battery positions. As it was, with greatly depleted numbers and with the necessity of guarding our left flank, we had to adopt a defensive attitude on our final objectives.

All the B Company officers had become casualties, and the sergeant-majors of both B and D Companies. Capt. J. C. Orchardson, commanding D Company, took command of the whole garrison, and he made the dispositions for defence, which successfully held the positions during the day.

As a result of their advance the East Surrey Regiment had an exposed southern flank. A company of the Guards were to come up to protect this flank, but until they came this was obviously a danger point, and there was no certainty that the Guards would be able to get through. From about 10 a.m. the enemy commenced to shell the village. The fury of his concentrated bombardment from then onwards beat almost anything we could remember in the way of shelling, with the possible exception of the attack on Delville Wood in 1916. Eight-inch and 5·9-inch shells were falling in the village at the rate of from six to ten a minute, and it became impossible to move troops anywhere in the streets. We had no troops ourselves in the village, except Battalion Headquarters and the Dressing Station, both of which had a most unpleasant day. The Guards' Company on their way up shortly after 1 p.m. got caught in the shelling, suffered very heavy casualties, and were unable to get to their positions. There was thus an unprotected flank on the right of the East Surrey positions.

A lot of gas shell was mixed with the other shelling, and this, in spite of the fact that there were still several hundred civilians in the village—a fact well known to the enemy. Months afterwards we visited Haussy, and were told on all sides of scores of civilians who had died from the effects of the gas on October 16. In the early afternoon, after the continuous shelling for over four hours, the enemy counter-attacked. He succeeded in penetrating the exposed flank south of the village, and he advanced as far as the river, some 300 yards south of the church. From this point, under the cover of an intense barrage, he worked through the village on the eastern bank of the river, and large numbers of the 9th East Surrey Regiment, holding positions on the eastern outskirts of the village, were thus cut off, and our situation north of the village became critical. Capt. Orchardson was ordered to take up the line of the sunken road from the chapel to the railway, and to form a strong post at the chapel facing south, to deny the enemy the northern exits from the village. All this was done promptly, and just in time, and the enemy made no further progress. In the midst of all this inferno we were informed that two companies of the 7th Northamptonshire Regiment were on their way up to relieve us. It was a precarious position to hand over to them. We were the only troops left holding positions on the east of the river, but it was still a fact that we were able to hand over to them at the close of the day the whole 600 yards of river and bridge-heads captured by us in the early morning. During the day also we had added one or two improvised bridges, and the position we left was a very valuable one.

We feel that the Battalion can be justly proud of its share in the Battle of Haussy. The very daring and successful work of A Company, in cutting out part of the village two days before the attack and holding it, had a most important bearing on the operations. On the day of the attack the work of the attacking companies was beyond praise. In spite of very heavy casualties at the outset, all ranks displayed the greatest determination. The fact that all objectives were taken, together with a large number of prisoners and a great deal of material, reflects the very greatest credit on Capt. Orchardson and the other officers, N.C.O.s, and men. Throughout the very trying day which followed they held

THE MEMORIAL BOARD AT HAUSSY.

THE RIVER SELLE AND THE MILL FROM WHICH MANY PRISONERS WERE TAKEN.

their ground with the greatest tenacity, in spite of repeated enemy attempts to dislodge them, and the crossings of the river were handed over intact at the close of the day.

At the chapel is a cross erected to the memory of our men who were killed in the Battle of Haussy. 2nd Lieuts. Laskey, Bain, Holman, and twelve other ranks were killed, 2nd Lieut. Wallis and Coy.-Sergt.-Maj. Gutteridge died of wounds, and Capt. Porter, Coy.-Sergt.-Maj. Dyer, and seventy-five other ranks were wounded.

The following decorations were awarded to the Battalion for the Battle of Haussy and the period preceding it :

The Military Cross

Capt. J. C. Orcharson, Capt. A. J. Porter, Capt. J. H. Selfe.

The Distinguished Conduct Medal

Coy.-Sergt.-Maj. J. Gutteridge, Cpl. H. Stoneham, Pte. A. Woodmore.

Bar to the Military Medal

Pte. E. Bowes.

The Military Medal

L.-Cpl. F. H. Apps, L.-Cpl. J. Bellamy, Pte. C. R. Brown, Sergt. H. H. Cater, Coy.-Sergt.-Maj. C. R. Dyer, Pte. A. French, Pte. H. Lang, Pte. A. Miller, Sergt. W. Mitchell, D.C.M., Pte. H. Pipler, L.-Sergt. R. Taylor.

The Meritorious Service Medal

Sergt. H. R. Swain.

After relief on October 16 we moved back to billets at Avesnes, and the next day we marched to Cambrai, where we remained until the 26th, cleaning, refitting, and training. Cambrai was still in an uncertain state. Parts of it were liable to be blown up by delay-action mines. During our stay there the railway lines at and near the station were constantly blowing up. The houses, even if undamaged outside, were wrecked inside. Furniture, pictures, books, and crockery had been broken or mutilated very thoroughly and wantonly in almost every house by the enemy before he left. Our first work was to try to get the houses we were to use as billets into something like order.

A large part of the town—the part which had been burnt—was barricaded off, and no one was permitted to explore. After a day or two we had got our billets into shape, and, having become accustomed to the constant explosions in the town, we had a comfortable time.

While we were at Cambrai the Quartermaster went into hospital after struggling gamely with very indifferent health throughout the whole summer, and there can be no doubt that the loss of Capt. Wood was more keenly felt by all in the Battalion than would have been the case with any other individual. In the opinion of everyone who served with him, he was the finest quartermaster who ever issued a rum ration. He joined the Battalion early in 1916, and for more than two and half years he worked untiringly for the welfare of the men.

Nearly every officer who passed through the Battalion will remember the kindly hospitality he received at one time or another at the Quartermaster's Mess, and Wood had the happy knack of making all feel at home at once. One of his great troubles was getting the stores loaded on to the baggage wagons for a move. Almost invariably, after eyeing the amount to be taken, and with a particularly severe glance at the size of the officers' kits, he would say, " I can't carry it," but, as he always did succeed in carrying it, this became a standing joke. His work may be summed up by saying that he was the friend of everyone in the Battalion.

On October 26 we returned to Avesnes, where we remained till November 2. On November 2 we moved up to Haussy, now some distance behind our advanced guards, and perfectly peaceful. We spent one night there. We met again many of the civilians who had received us so well and helped us during the fight. They were greatly rejoiced at being free of the Boche, but they were perturbed at the damage done to their village, and very bitter about the enemy's use of gas shells on October 16. During the night the enemy threw over three or four shells from very long range.

On November 4 the 24th Division relieved the 61st Division. The 73rd Brigade formed the advanced guard, the 17th Brigade were in support, and the 72nd Brigade were in reserve. We moved up to Sepmeries, where we were billeted for the night.

The nature of the fighting had changed considerably since we were last in it. Up to the Battle of Haussy the enemy had been putting up a stubborn resistance, especially where he was defended by a natural obstacle such as the River Selle. Up to then, when there was a fight, it was a very stiff affair, and counter-attacks in great force, supported by annihilating artillery fire, were always likely.

Now it was a case of taking up the hunt of a really retreating enemy. We would still look for fights and we would expect to be shelled at intervals, but the fights were likely to be skirmishes with rear guards of machine gunners rather than with the main enemy force, and the shelling was not likely to be concentrated and continuous as it had been at Haussy. We were to find the advance trying enough, however, for the distances we were to be called on to cover each day were very great, and often in unexpected quarters a company or a platoon would get involved in a pretty desperate local encounter.

However, we felt we were now really entering into the open fighting we had hoped for all through the long years of the war. Although we were still in Divisional Reserve, we could realise by the length of our marches each day what sort of thing was going on in front of us.

On November 5 we marched again forward to Wargnies-le-Grand, where we remained in billets for two nights. Rain was almost continuous while we were there, but the whole Battalion was in excellent spirits. Our long stay at Wargnies was on account of a hold-up on another river line at St. Waast-la-Vallée in front. The advanced guard had not succeeded in taking this in its stride, and an attack was therefore arranged for November 7 at 6 a.m.

The 17th Brigade were to take St. Waast and stop there, and the 72nd Brigade were to advance through them and take up the advanced guard work. At 4 a.m. on the 7th we left Wargnies, and our Battalion remained the advanced guard of the Brigade for the rest of the war.

So little news had penetrated through to us that we had no idea as we marched off in the pitch blackness of that November morning that we were in the last days of the war. We had begun to feel that we should advance on and on to the Meuse, then to the Rhine, then across the Rhine, and then anywhere—to Berlin perhaps. The idea of an armistice had never occurred to us. The village of St. Waast was found to be evacuated by the enemy, and we passed through the Rifle Brigade on to the high ground running north-west from Bavai. The Guards entered Bavai from the south at the same time, and we continued our advance during the day to positions some 2,000 yards east of Bavai. Here we were met by very heavy machine-gun fire, and any advance beyond this line was found to be impossible, and we therefore dug in for the night and pushed patrols forward. Strenuous efforts were made by these patrols during the night to cut out some of the enemy's machine-gun positions, but he maintained a very heavy fire throughout the night from positions which had obviously been carefully chosen beforehand, and dug by the unfortunate French civilians of the district. Lieut. H. R. Smith and 2nd Lieut. Everson were killed at St. Waast at the commencement of the advance.

At Bavai on this day and the next we attached to ourselves various units, some officially sent to us, some found by the wayside. In this way our Battalion for the rest of the war became a small field force, complete in almost every branch of the service. The force consisted of:

8th Royal West Kent Regiment (ourselves), 6th Dragoon Guards (found at Bavai, and glad to come along with us), one section of 18-pounder guns (sent to us officially), one section of 4·5-inch howitzers (found, and keen to remain with us), two sections machine gunners (sent to us officially), one section Stokes Mortar Battery (sent to us officially), one section 6-inch Newtons, under command of our old friend Lieut. Bennett of Hargicourt days. He had brought up his section complete, looking for work, and was delighted to come with us. How all these units were fed during these last days remains a mystery, but we had a splendid time together, and we certainly did not starve.

At 6 a.m. on Friday, November 8, we continued the advance with the North Staffordshire Regiment on our right. The 6th Dragoon Guards did our scouting for us during the whole day, and fairly cleared the ground of the enemy.

By 2 p.m. we had advanced 7,000 yards or

more over difficult wooded country. It was great fun, but very tiring. The work of the junior N.C.O.s was particularly good. A lance-corporal with his section of six men would constantly find himself advancing entirely out of sight of the sections on either side. As far as possible we had provided every section commander with a sketch map, and the way they kept direction was really splendid.

On a line some 3,000 yards west of Feignies Station and the village of La Rolies the enemy was again found in very strong positions, and very heavy machine-gun fire was again encountered.

We got up the 6-inch Newtons and the Stokes, and these, with the assistance of the 18-pounders and the 4·5-inch howitzers, shot up the likely places for machine guns. We dislodged a number, and the civilians at La Rolies next day told us of casualties being carried through the village during the night. We advanced a little further before nightfall, and then found ourselves fairly up against it. The enemy machine gunners evidently had instructions to get rid of all their ammunition, and they shot almost without ceasing and made the work of our patrols very difficult. It was bitterly cold, and every man in the Battalion was absolutely beat. We had received orders during the day that we were to stop at night, and other troops would pass through us to take up the chase next day. We had therefore put absolutely every ounce we knew into it that day.

Then at 11 o'clock at night our Commanding Officer was called to Brigade Headquarters to a council of war, in which orders were read to the effect that the 72nd Brigade would continue the advance, and the 8th Royal West Kent Regiment would again lead the advanced guard. It is reported that our colonel, who was almost asleep in his chair, on hearing the orders, uttered two words, "Good God!" and fell asleep with his head on his hands. This was Gen. Morgan's story, and it was vouched for staunchly by his staff afterwards, but we can hardly believe it. Still it gives some idea of the fatigue which had almost overcome everyone of us on this night of November 8. The pace had been very hot indeed for nearly a week, and on this last day, believing it to be our final go for a time, we had put in that extra spurt which had left us pretty limp.

So on November 9 we were to take Feignies Station and the village of La Rolies, stop there, and let the rest of the Brigade—and we hoped the rest of the British Army—pass through us.

Col. Wenyon tells us that he was most distracted when he got back from that Brigade Conference. He did not know if the men could march the two and half miles to Feignies Station, and, if they could, they certainly could not be expected to fight much, and there was a nasty looking embanked railway to cross.

It was decided finally that the patrols which were out already should be instructed to push on immediately it was light and find out whether there was likely to be a fight. A message was sent to our friends the Cavalry, who were only too pleased to assist. So soon after dawn the mounted men set off. Two platoons, one from D Company, under Lieut. Gedge, one from B Company, under Lieut. Michie, continued the advance they had made during the night, followed by our Commanding Officer and his groom. It was a close finish. The Cavalry men just got there first. Lieut. Gedge, unshaven and dirty, looking like a backwoodsman, was just extricating himself from his tenth embrace by the worthies of the village of La Rolies, when Col. Wenyon arrived. Flags began to flutter from the windows, gallons of coffee were brought out in small cups from every house in the village, and a rusty old gramophone began to play the Marseillaise—the first time for over four years. The enemy had left on bicycles a few minutes before we got there! Our cavalry friends were after them, and got a few later on. Message was at once sent back to the rest of the Battalion that all was well, and they could come quietly up into billets. During the next two hours they all arrived, and were received with the greatest jubilation on all sides. Coffee was poured down throats, cauldrons of hot water were produced, and sore feet and aching limbs were washed and bathed in a most reverent manner by the admiring civilians. Battalion Headquarters was established in the Hôtel du Nord, opposite the demolished station. The rusty gramophone continued to play the Marseillaise all day, the village ladies donned such finery as they had left to them, and strutted the village street, trumpets were blown, more flags were produced, and with laughter and tears the men and women of that little village passed from war to peace. These people had seen the first great onrush of the German Army in 1914—for Maubeuge was only a stone's throw away—they had borne the German tyranny over four years, the railway station had been burnt to the ground during the last week, and now, almost before the hated enemy had beaten it out of the village, *voila les Anglais—et liberté.*

So ended the fighting of our Battalion in the Great War. That night we heard of the abdication of the Kaiser and the offer of the Armistice.

We slept long and comfortably, almost every man in a proper bed. We began to feel somehow that this was the end. It was more than a trifle disappointing to the soldier in the field of course. We wanted to see the German Army, if not annihilated, at least squarely beaten in a fight. And it looked as though he was to get away without his whipping.

If this was *not* the end, we were ready to take up the chase again, and run after the Hun to the end of the world; if it *was* the end, we had a notion that in our small corner of this world-wide four years' struggle we had tried to do our job.

We must mention that at Feignies Station on this last day we captured an enemy minenwerfer factory complete. It was an enormous place, with sufficient ammunition to blow up most of the German Army, and we think we are probably safe in saying the most valuable single capture made in a single day by any battalion in the war. It was a pity it could not have been captured years earlier.

During the whole of this final advance from Niergnies to the Mons–Maubeuge road the 24th Division advanced during the six weeks a distance of fifty miles. During the last three days, from Wargnies-le-Grand to Feignies, our Battalion led the advanced guard over seventeen miles.

During the six weeks the Battalion captured 250 prisoners, two field guns, three tanks, one enemy minenwerfer factory, and large numbers of dumps of all kinds.

Our casualties were 9 officers and 48 other ranks killed or died of wounds, and 5 officers and 113 other ranks wounded.

PHASE X

FROM THE ARMISTICE TO THE DISBANDING OF THE CADRE

November 1918—*June* 1919

THE early morning of November 11 found the Battalion still at La Rolies, under the command of Maj. R. G. Rogers, the Commanding Officer and 2nd-in-command both being on leave. At 5 a.m. on that date, what was to be the final march back from the forward area was commenced, and the Battalion passed through Bavai to Pissotiau, a distance of some seven miles. On the route the news of the signing of the Armistice was read by a D.A.P.M. Passing Bavai Station a group of A.S.C. men asked what the cheering was about, and were told by one of our men that the infantry in the line were tired out and were being relieved by the A.S.C.

The billets at Pissotiau were very bad and overcrowded, and it is small wonder that the news of the cessation of hostilities, which arrived about 1 p.m., evoked no great enthusiasm. To make matters worse, there was no beer or other suitable beverage with which to celebrate it. There is a rumour that one officer found a partially filled rum jar and had a small celebration all to himself, but we take no responsibility for the story, and give it for what it is worth.

The following day the accommodation was slightly improved. The inhabitants gave us a warm welcome, and were in no way to blame for our discomforts.

The ordinary routine of life out of the line was carried on, including on the 13th a much-needed visit to the baths and an issue of clean clothing. On the 14th a thanksgiving service was held for the cessation of hostilities.

On the 17th we left Pissotiau without regret, and marched, hoping for better billets, about eleven miles to Sepmeries, but found them very nearly as bad as those we left. The weather for some days had been extremely cold, and the Battalion had little cause to enthuse over its first few days of peace.

On the 18th the march was continued to Lourches, where good billets were found and much appreciated. As the crow flies, the distance between Sepmeries and Lourches is about eleven miles, but many roads were impassable on account of destroyed bridges, and the distance covered was very much greater. In addition to other discomforts, the men's boots were in very bad condition, as there had been no fresh issue since the heavy work of the final advance.

The next day a further move, this time quite short, was made to Bruille.

Throughout the march the most noticeable thing was the large number of French refugees returning from the occupied territory.

All British and French lorries travelling in a westerly direction were full of them, and they were making their own way in streams along each of the main roads, getting a lift here or there from a staff car, lorry, or transport wagon, and then struggling bravely on on foot, many bootless and still more hungry.

On November 20 Maj. Brown arrived back from leave and took over command, and a couple of days later Maj. Rogers went into hospital with trench fever and did not rejoin us. He commanded B Company for nearly nine months, and was a most successful company commander, very reliable and extremely thorough. Lieut. H. G. Michie took over the company for a few weeks, and then Capt. A. Drumgold, D.S.O., was transferred from C Company to command permanently. A few days previously, Capt. C. F. Penton, who commanded D Company from immediately after the Battle of Loos until he was wounded on the Somme, returned and resumed command of his old company.

On the 21st we moved a couple of miles to Masny in order to get better accommodation, and the march to winter quarters was continued on the 25th to Landas, about ten miles, and again the next day, when we had a short journey to Nomain.

Noma n was a large village which accommodated ourselves and the North Staffords. It was very similar to any other French village of the same size, and afforded fairly comfortable billets without being in any way luxurious. It had the advantage of containing quite a number of passable football grounds, which were, as a rule, very scarce in this part of the world. Each company had its own ground and made full use of it. Boxing and running were also taken in hand, but never attracted the same popularity as football.

Originally we expected to stay at Nomain for a few days only, and then to move on to Tournai, but the troops already there, who were to proceed to Brussels, were delayed from one cause and another, and eventually we only got into Tournai just in time to make arrangements for Christmas.

While at Nomain we twice had lorries to take some of the Battalion over to Lille for a joy ride.

Perhaps the outstanding event of the stay at Nomain was the mock election which was carried on with great enthusiasm for three days. It was at the time of the General Election at home, and, in order to place the issues before the Battalion, the Colonel, who had returned from leave on November 29th, selected five candidates, each of whom had to undertake to stick to the official programme of the party he was to represent. The whole scheme was carried through in the spirit in which it was started, and no attempt was subsequently made to take advantage of the relaxation of certain points of discipline which was necessary during its progress. The candidates were: Coalition, Cpl. Wray; Independent Liberal, Pte. Daft; Labour, Capt. Selfe; Independent (John Bull Party), Lieut. Jackson, and Discharged Soldiers and Sailors, Pte. Airs.

The campaign was started by a general meeting, at which each candidate had ten minutes in which to lay his programme before the constituents, and then ten minutes answering questions, after which each one carried on his campaign separately. The issue soon developed into a fight between Coalition and Labour, and eventually the Labour candidate just got home, largely, we believe, though we may be prejudiced, by frequent allusions to the working-man's beer.

General Morgan attended one meeting and spoke in favour of the Coalition candidate, who also had the support of Lieut. Fitch, the educational officer; the latter gave an impassioned harangue of three-quarters of an hour in broad Canadian dialect, which should have finally convinced any waverers of the justice of his cause.

There can be no doubt that the great majority of us knew a lot more about the issues before the electors at home at the end of those three days than we did at the beginning. The balloting took place on December 12, and in the evening a Battalion Concert was held in the school building, at which the progress of the count was announced. It was at this concert that the news was received of the award to Lieut. D. J. Dean of the Victoria Cross, won two and half months before. The scene at this announcement has been described in an earlier chapter.

On December 7 His Majesty the King visited Tournai, and eighty-five officers and other ranks were taken by lorry to represent the Battalion on the route from Tournai to Lille.

On December 3 the advanced party, which had been at Tournai since our arrival at Nomain, returned, the move being indefinitely postponed. Only the billeting officer, 2nd Lieut. W. E. W. Thomas, remained to look after our interests. The barracks at which the troops were to be quartered were definitely allotted to the Brigade, but the areas for officers' billets and messes were not quite so clearly defined, and there was a distinct tendency on the part of the unscrupulous to snatch the best accommodation available. 2nd Lieut. Thomas succeeded quite well in resisting the attempts of these marauders, but even then some of the company messes and billets were not as good as they should have been for times of peace, and on the day of arrival at least one mess had been annexed by a Divisional Office for its junior clerk!

It was at about this time that we heard of the award of the Croix de Guerre by the President of the French Republic to 2nd Lieut. Thomas, for assistance in the evacuation, under heavy fire, of the civil population from Haussy in October. His perfect knowledge of French was frequently of great service to us.

Fresh advanced parties proceeded to Tournai on the 17th, and the Battalion followed the next day, arriving in time for dinners. The town had been evacuated by the Germans without a fight. The outskirts showed signs of some shelling, and the bridges over the Scheldt had been destroyed by the enemy, but the town itself had been very little damaged except for the railway station and its neighbourhood, which showed every sign of the effectiveness of our aerial bombing. The cathedral and the main square seemed to be quite untouched. The population was distinctly friendly, as was usually the case in all districts which had been in the hands of the enemy. All the shops were open, and they had succeeded in collecting very fair stocks.

The barracks consisted of five massive buildings round an open space or barrack square. Each battalion had one of the buildings and a portion of a fourth for its quarters, while the fifth was divided up for use as class-rooms under the educational scheme and for canteens and recreation rooms. We do not doubt that the Germans left the barracks in a state of absolute filth, but at the same time British troops had occupied them for some weeks, and we do not think that there is any excuse for the awful condition in which they were when handed over to us.

It is difficult to see that any effort at all had been made to clear away the rubbish or clean the place generally, and it was in a very similar condition to the billets at Cambrai, which we had taken over direct from the Boches.

Needless to say, we very soon made enormous improvements all round, and in a short time had made the barracks as comfortable as the material available would allow. In the barrack rooms the stoves and windows were repaired, the rubbish cleared, and the whole place scrubbed out; broken bedsteads were repaired where possible, and fresh straw obtained for the palliasses. A sergeants' mess and company mess rooms were fitted up, and proper storerooms made for the Quartermasters' stores. The recreation room was taken in hand, furnished comfortably, and a piano obtained; newspapers and periodicals were provided as far as possible, and a lending library started with such books as could be collected or obtained from home. A ping-pong table was put up, and proved a great attraction, and a wet and dry canteen, including a coffee bar, was run in the adjoining room. The star turn was perhaps the barber's shop, which was fitted up with all modern conveniences, and where one could get a hair cut and shampoo little inferior to those obtainable in the West End of London. The Divisional Commander, when he came round to inspect the barracks, was particularly struck by the barber's shop, and remarked to the Brigadier, "This is the Battalion to get things done."

The company officers' messes and billets were passable. Battalion Headquarters, on the other hand, were most luxuriously housed. 21, Rue de L'Esplanade, was a house containing a "dining-room, two other reception rooms, conservatory lounge, six bedrooms, bath (h. & c.), and usual offices, company's water, electric light and gas," and the whole furnished with a comfort which would have done credit to an English home, and which was most unusual for the continent.

At about 6 p.m. on December 21 the alarming report was brought to Battalion Headquarters that the barracks were on fire, but on investigation it proved to be a building in close proximity, part of which was being used as a hospital, and the remainder—the part which was on fire—as billets for the 7th Northants, who had only come into the town and taken up their quarters there that afternoon. The corridor connecting the hospital with the burning building was quickly demolished by volunteers, mainly from the 72nd Brigade, and the former was saved, but the latter was absolutely gutted, in spite of the efforts of the local native fire brigade. It was a large building, and fanned by a good breeze the fire raged fiercely for about three hours, and was quite an impressive sight.

On arrival at Tournai, preparations for Christmas were taken in hand. As far as the dinners were concerned, we had been told that we could rely entirely on the Divisional Canteen, but, fortunately as it turned out, we decided to get what we could on our own, and the only thing which was left was the most important of all—the beer. This we were most solemnly assured would turn up without fail in time for Christmas, but on the morning of the 24th there was still no news of it, and on representations being made to the Brigade, they realised that the matter was serious, and obtained a lorry for each battalion to fetch its own. Lieut. Outram, who went in charge of our lorry, made straight for our old haunts round Bully-Grenay, where he knew all the breweries, and arrived back in the very nick of time with a good supply. It afterwards transpired that the truck containing the Christmas fare for our division got mislaid on the railway. All our company dining-halls were decorated, and the dinners of roast pork and beef, with vegetables, plum pudding, fruit, cigars, and at last the belated beer, were a great success, but tragedy was only just averted. The beer actually did not arrive until the dinners had been in progress for half an hour, and only those who have spent a Christmas on active service will realise what the absence of beer would mean. As usual the officers and sergeants waited at table.

The dinners were followed by equally substantial teas, and a Battalion concert, at which the star turn was a short sketch, written by the Commanding Officer and performed by the officers.

The sketch, entitled "How they spent it," was most amusing, but, without any aid from the book, the mere sight of the performers in the most ridiculous civil attire, borrowed from the local inhabitants, was enough to send the house into roars of laughter; Mr. Bowen, attired in tall hat and tail coat, both much too small, over khaki trousers, and with a very high collar, was perhaps the funniest of the lot. The principal characters were: an old gentleman, Mr. Fitch; his solicitor, Mr. Thomas; his sisters, Capts. Penton and Mackaskie; his brother-in-law, Mr. Bowen; his niece, Mr. Binney; and his profligate nephew "Archie," Capt. Drumgold. The acting of the two sisters was really quite clever, and the sketch was received with shouts of merriment which drowned a good deal of the dialogue.

On Boxing Day, in vile foggy weather and in a sea of mud, the rest of the Battalion beat the officers at football by 3–1. It was a very poor match on account of the weather conditions, but was designed mainly to work off the effects of Christmas, and was partially successful.

The great thing in these days was education. Lieut. Fitch, who was appointed educational officer, carried out his duties with the greatest enthusiasm and energy, and made the scheme a success in the face of many difficulties. It was decided that the only practical way of running the scheme was to make the platoon the unit, and the platoon commander responsible for the education of his command just as he was for its military training. The educational officer is to be congratulated on the results. The scheme of education by platoons was originated in our Battalion, and afterwards adopted throughout the Division.

In connection with the educational scheme, a Dramatic Society was formed. This was quite voluntary, and attracted many members, who met in the evenings for the purpose of reading Shakespearean or other plays. The members were divided into groups, and each group chose its own play. One of the groups intended producing an abridged version of "Julius Cæsar," but the order for a large draft to proceed to the Rhine cut short the project.

A special group was formed for the production of Arnold Bennett's Comedy, "The Title," which was having an excellent run at the Royalty Theatre in London. Lieut. Fitch wrote to Mr.

Arnold Bennett, who gave permission for the production of the play, provided that any proceeds went to charity, and who also sent us a copy of the book.

The reading and rehearsing of the piece was commenced on December 30, and lasted for nearly five weeks, during which a great deal of real hard work was put in. Col. Wenyon was producer, and the play was originally cast as follows :

Mr. Culver		LIEUT.-COL. H. J. WENYON.
Mrs. Culver		LIEUT. B. HIGGINS.
Hildegarde Culver	their children	LIEUT. R. C. C. J. BINNEY.
John Culver	their children	LIEUT. F. C. M. CHAUNCY.
Tranto		MAJ. H. S. BROWN.
Miss Starkey		LIEUT. J. E. JORDAN.
Sampson Straight		LIEUT. W. E. W. THOMAS.
Parlour Maid		LIEUT. H. R. JAMES.

Two changes had subsequently to be made, which caused a considerable set-back in the progress of rehearsals. Maj. Brown left early in January to attend a course at Aldershot, and Lieut. Thomas was claimed by the Brigadier for an important part in the revue being produced by the Brigade Concert Party. Lieut. E. O. Goulden took over the part of Tranto, and Capt. J. H. Selfe succeeded Lieut. Thomas as Sampson Straight.

The play was produced in the salons at Battalion Headquarters, 21, Rue de L'Esplanade, on the evenings of February 1, 3, and 4, 1919.

The production was a great success, and well repaid the thoroughness of the work which was given to it. The whole of the cast did well, the best performances being perhaps by Col. Wenyon and Lieut. Chauncy.

On the first night most of the battalions and other units in the Division were represented in the audience; the second performance was attended by the Divisional and Brigade Commanders and their staffs, and the third performance, which was followed by a dance, was honoured with the presence of the nurses from the British Hospital.

Another dramatic effort was the revue produced at the Tournai Theatre, under the personal direction of Gen. Morgan, by the Brigade Concert Party, and called "Harem Scarem," or "The Sultry Sultana." It was originally intended as a Christmas entertainment, but did not appear until much later. The delay was partly caused by the funny man, a really excellent low comedian, carrying his humour to the point of getting himself demobilised without warning when at home on special leave to procure properties for the production. As he had one of the longest parts, and rehearsals were at the time well advanced, his behaviour caused a serious set-back, and 2nd Lieut. W. E. W. Thomas was roped in by the General to deputise. In the first act, which represented Diogenes and the Tub, Gen. Morgan himself appeared most successfully as Diogenes.

As was to be expected with a production under Gen. Morgan, the revue was very well staged, and had a successful two weeks' run.

The Divisional Concert Party also appeared

at the theatre, and there were cinema shows in the town to help the troops pass their evenings. Dancing became very popular with many of the officers, C Company being pioneers in this respect.

In the earlier stages of the stay at Tournai, training and other work continued very much as usual, except for the substitution of education for military training during a portion of the time available, but as demobilisation advanced and demands for working parties increased, the number of men available for parade gradually diminished, and after a time every man had to be used on some form of employment or fatigue.

So much looting from supply trains had been going on, that D Company was permanently billeted in the vicinity of the station for the purpose of picketing all trains that came in.

Football was, of course, the principal sport. On arrival we had the usual difficulty in finding a suitable ground, but we eventually secured one on the outskirts of the town some little way from the barracks, which, after sundry shell-holes had been filled in and turfed, proved to be extremely useful. An inter-platoon competition was held in the Battalion, an inter-company competition in the Brigade, and an inter-unit competition within the Division. Our Battalion team was a very useful one. Coy.-Quartermaster Sergt. Judd in goal, Sergt. Scott at centre-forward, and the Medical Officer, Capt. Frier, among the half-backs being perhaps the most prominent. In the divisional competition we were beaten in the semi-final by the attached army brigade of artillery, who were a very hot lot, and who, we believe we are right in saying, eventually won the competition.

The Company Competition was never finished, and the final of the Platoon Competition between (we believe) Nos. 1 and 13 Platoons ended in a draw and was never replayed.

Joy rides by motor-lorry became fairly frequent in the later stages at Tournai. Sometimes these meant a visit to Lille, and at others a trip to some part of the old battlefields at which the Battalion had fought, and they included visits to the Salient, the Lens area, the Somme, and a round trip via Cambrai, Awoingt, Haussy, and Hargicourt to St. Quentin, and back along the route of the March Retreat to Amiens.

Demobilisation commenced on Jan. 2, and went on steadily until the end of February. In January 113 other ranks were demobilised, and in February four officers and 221 other ranks. On February 6 a warning order was received for ten officers and 200 other ranks to be prepared to join the 10th Battalion in the Army of Occupation on the Rhine, and on February 25 the four-company system was abandoned and the Battalion was reorganised into a draft company under Capt. Penton, and a cadre company under Capt. Thompson. By the end of February demobilisation was temporarily complete.

On March 2 five riding horses were transferred to the 9th East Surreys, who were joining the Rhine Army, fifteen other animals went off by train for demobilisation, and a further five left to be sold locally; the next day a further fifteen were demobilised.

A spell of very cold weather was experienced at Tournai, starting with a heavy snowfall on January 26 and 27, and the snow did not finally disappear until February 17.

On February 21 the ceremony of the presentation of the King's Colour took place, and to enable the Battalion to practise the necessary ceremonial all our duties were taken over by the 3rd Rifle Brigade from the 16th to the 22nd.

The ceremony took place on the Champs de Manouvres, and the troops on parade were the 9th Battalion East Surrey Regiment, the 8th Battalion The Queen's Own (Royal West Kent Regiment), under the command of Capt. C. F. Penton (Lieut.-Col. H. J. Wenyon, D.S.O., being unfortunately unable to attend), and the 12th Battalion Sherwood Foresters, the parade being under the command of Brig.-Gen. R. W. Morgan, C.M.G., D.S.O. The presentation was to be have been made by the Army Commander, Gen. Sir Henry Horne, K.C.B., K.C.M.G., but he was unfortunately indisposed, and deputed Lieut.-Gen. Sir Arthur Holland, K.C.B., D.S.O., commanding 1st Corps, to take his place.

The Brigade was drawn up on three sides of a square, with the Queen's Own in the centre. The weather, which had been very wet during the night and early morning, was brilliantly fine at 11 a.m. when Sir Arthur Holland arrived, accompanied by Maj.-Gen. A. C. Daly, C.B., commanding 24th Division and the Corps and Divisional Staffs.

The Corps Commander was received with the General Salute, and the consecration service then commenced. On its conclusion the Colours were handed to the Presenting Officer by the Seconds-in-Command, and were received from him by the Senior Subalterns kneeling on the right knee. Capt. F. Proctor was acting as our 2nd-in-command, and the senior subaltern was Lieut E. O. Goulden, M.C. The Colours were then marched to their battalions, and received with the Royal Salute.

The Presenting Officer then addressed the parade as follows :

" Brig.-Gen. Morgan, Officers, Warrant Officers, Non-Commissioned Officers, and Men of the 72nd Infantry Brigade. The first thing I have to do is to apologise for the absence of the Army Commander. He had fully intended to come, and had been looking forward to the privilege of presenting these colours, but I am sorry to say he is unwell, and consequently has been unable to come. I have, therefore, the privilege of being here in his place, and the honour of presenting the King's Colour to the battalions now on parade, an honour enhanced in this case by the fact that the presentation takes place at the moment of victory, and to battalions which have fought in this the greatest war in the history of the British Empire.

" The Colours which I have entrusted to your keeping are the symbols of your allegiance to your King and to your Country, and as such they will, I know full well, be honoured and guarded by you on all occasions and at any price, and that you by your deeds and by your conduct will worthily uphold the glorious traditions of your regiments, not only in war, but equally in peace. I want you to remember that the trials which will face you in peace are as great and more insidious than the trials which you have so heroically overcome during this war, but if, when you return to civil life, you face these trials with the same loyalty, steadfastness, and self-sacrifice which have been the mainsprings of your actions and the foundations of your success out here, then I am convinced that no stain will be placed on the fair fame of your regiments by any act of yours.

"The 72nd Infantry Brigade has a magnificent record throughout this war. It first fought at Loos in 1915, after a long and trying march up from the Coast. It was next engaged in the attack on Guillemont and in the heroic defence of Delville Wood. It was then transferred to the 1st Corps, and performed fine service during the Vimy Ridge operations, when just prior to the attack it held the Canadian assault trenches, and endured with the utmost steadiness the brunt of the German artillery counter-preparation. The Brigade then moved north, and took a distinguished part in the Battle of Messines and in the Third Battle of Ypres. In the Great German Offensive of 1918, the 72nd Infantry Brigade played a glorious part, holding up many overwhelming attacks, and by its courage and fighting spirit saving the situation on many occasions.

"To such gallant battalions I, with absolute confidence, entrust these Colours in the name of His Majesty the King, asking you to remember that, as they represent all that is most glorious in the past history of your regiments, so they must be an inspiration enabling you to add to your Country's glory in the future."

Brig.-Gen. Morgan, in a brief speech of thanks, said:

"Sir,—In the name of the 72nd Infantry Brigade, and on behalf of the battalions on parade to-day, I beg to thank you for the honour done to them in the presentation of Colours, and for the reference to their services in your address. These battalions are justly proud of their record in the war, and, speaking for myself, I have always found it a very great honour to have them under my command.

"All may soon cease to exist as units, but the Colours will remain as a lasting symbol of their loyalty and devotion to His Majesty; and, should it ever be found necessary to reform the battalions, the flags which you have just presented to them will form a rallying-point round which will again be displayed that lofty sense of duty, comradeship, and devotion which has ever distinguished the 9th Battalion East Surrey Regiment, the 8th Battalion Royal West Kent Regiment, and the 12th Battalion Sherwood Foresters."

Battalions then moved to the Saluting Base, and marched past, after which the Brigade formed up in line, advanced in Review Order, and gave a General Salute.

Sir Arthur Holland congratulated all ranks on their turn-out and steadiness in the ranks. The parade was then dismissed.

On March 4 the 9th East Surrey Regiment, which had been with the Brigade since its formation in September 1914, left to join the Rhine Army.

On March 5 orders were received for the reduction of the establishment to Cadre A, and on March 16 four officers and two other ranks were demobilised, followed on the 30th by one officer and nineteen other ranks.

The draft for the 10th Battalion left on March 17, and consisted of Capt. Drumgold, Lieuts. James, Bowen, Binney, and Beynes, and 131 other ranks. Capt. Penton, Lieuts. Brunger, Jackson, Gibbons, and Smith, with the balance of the men, proceeded to the Rhine on their return from leave. Lieuts. Gobell, Herd, and Lane, with twenty-nine other ranks, also went to the 10th Battalion on April 10, and a further ten other ranks on the 22nd. Three officers were demobilised on April 16, and twelve other ranks on April 24, and this practically reduced the Battalion to the revised establishment for the permanent Cadre.

Before leaving Tournai a farewell ball was given by the N.C.O.s of the 72nd Infantry Brigade to the inhabitants of the town. It was a great success, and nearly all the officers and N.C.O.s of the Brigade were present during the evening, the Brigadier, with a fair Tournaisienne, leading the grand march at the commencement of the second half of the programme.

On March 21 the barracks were evacuated, as Belgian troops were coming in, and the men were billeted in the Chaussée de Lille, Battalion Headquarters remaining at 21, Rue de L'Esplanade.

On March 26 the Cadre, less the transport who followed on the 28th, left Tournai and took up its quarters at Lamain, a village four miles west of the town, just off the main road to Lille. Lamain was a straggling and thoroughly rustic village, and the men were quite comfortably billeted by twos and threes about the farm-houses. The stay at Tournai had been very pleasant, and at first the troops did not much like the change to a rural life, but they soon settled down.

Joy rides by motor-lorry continued, one of these from Lamain being a three-day trip to Ostende, Zeebrugge, Bruges, and Blenkenberg. On the return journey the lorry broke down at Pithem, a village with no accommodation and some miles from anywhere. The station-master, on being approached, said that he knew a hotel proprietor in the next town who would be delighted to put up the party, and wired to this worthy to be prepared to receive them. On arrival they found the place, including the hotel, to be in ruins, but no British troops had previously been seen there, and their arrival was looked upon as a great event. Mine host, with many apologies for the circumstances which made it impossible to provide each visitor with a separate apartment, proceeded to do them very well indeed, and parted from them next day with many protestations of eternal friendship. The railway officials insisted on giving the whole party a free warrant home, and so the trip ended very satisfactorily.

This practically concludes the history of the Battalion overseas. The Cadre finally left Lamain for home, entraining at Baisieux in two parties on June 2 and 3. Lieut.-Col. Wenyon went home on leave separately, and Capt. Proctor was in command, the other officers being Capt. Mackaskie, the adjutant, 2nd Lieut. George, and Lieut. and Quartermaster Green, with Regt.-Quartermaster-Sergt. Dunk and some thirty-five other ranks. The party embarked at Antwerp on the 6th, and reached Tilbury on the morning of the 7th, after a fine crossing. They immediately proceeded by train to Prees Heath, near Whitchurch, in Shropshire, where Lieut.-Col. Wenyon rejoined, and there the transport was handed in and the whole Battalion finally demobilised.

A word here will not be out of place on the services to the Battalion of Lieut.-Col. H. J. Wenyon, D.S.O. He joined the Battalion as a platoon commander immediately after the Battle of Loos, and succeeded Capt. Wood in command of A Company at the end of May 1916. After six months attached to the staff he returned as 2nd-in-command in August 1917, and assumed the command on the departure of Lieut.-Col. N. I. Whitty, D.S.O., in December 1917, and continued to hold it until the disbanding of the Cadre in June 1919. His period of command included, in addition to the trench warfare in the devastated area in the early part of 1918 and at Lens in the summer, the fifteen days' continuous fighting which started with the great enemy attack on March 21, and also the final advance from Cambrai in October and November. In both these battles the success with which the Battalion fought was largely due to the untiring energy and splendid example of the Commanding Officer. He earned the D.S.O. at Inverness Copse in September 1917, and a bar to it for his great work during the March Retreat, and was subsequently awarded the Ordre de la Couronne and the Croix de Guerre by the King of the Belgians. Whatever he undertook, he carried through with a determination and energy which communicated itself to all ranks under him, and because he expected nothing but the best from everyone, he got the best.

It would be possible here to mention the names and services of many officers, warrant officers, N.C.O.s, and men which have not been specifically mentioned in this narrative, but if the list were once commenced, it would be difficult to know where to stop, and the inclusion of some names and exclusion of others might appear invidious. We feel, however, that the mention of two names will not be out of place.

Regt.-Sergt.-Maj. F. Greenaway, D.C.M., who

came out with the Battalion as a L.-Sergeant, continued (with one break of a few months during which he was at home on the sick list) to serve with it until after the Armistice. He did most gallant work on the Somme, and was awarded the D.C.M. for bravery during the Ypres fighting of 1917. He was a most popular Company-Sergeant-Major of A Company for over a year, and took up the duties of Regimental-Sergeant-Major when Regt.-Sergt.-Maj. Rankin was killed in May 1918.

Sergt. William Carr, M.S.M., also came out with the Battalion, and he never left it except for periods of short leave until he was demobilised in January 1919. He took over the duties of Orderly Room clerk immediately after the Battle of Loos, and was from then onwards a great mainstay to many successive adjutants. He knew the regimental number, name, initials, and history of nearly every man who ever served with the Battalion. On receiving a query from the Record Office *re* one, say, Pte. Jones, the Adjutant would inquire if Sergt. Carr remembered him, and the reply would probably be, " Yes, sir, No. 4286 Pte. Jones, P. He joined us at Messines, and was transferred to the Labour Corps when we were at Hulluch." A lion for work, he never spared himself under any circumstances, and his services to the Battalion can hardly be estimated. His Meritorious Service Medal was a very well-earned decoration.

The last act before the final demobilisation of the Cadre was the escorting of the King's Colour to the Regimental Depot at Maidstone on June 14, 1919.

The journey of the escort party of Non-Commissioned Officers, under the command of 2nd Lieut. F. H. George, M.M., was made with very little ceremony, and was unknown to all but a very few of the thousands who had served with the Battalion. Yet it was an important event, for at the moment when the King's Colour was finally handed over at Maidstone, the 8th Battalion The Queen's Own (Royal West Kent Regiment) was disbanded—as a military unit. Officers and men who had served with and loved the Battalion had marched out into their new walks of life, and were scattered through the length and breadth of the country and throughout the Empire. Yet among these thousands there is not one who does not feel that with its disbanding the Battalion has completed only its first great task, and that other duties lie ahead which will still test its powers to the utmost. In the war the real strength of a battalion could be gauged from the conduct of its men when detached from their unit. During the final advance some of the greatest work of all was that done by isolated parties of young soldiers under the command of junior non-commissioned officers, working often for many hours on difficult patrol work, without any possibility either of keeping touch with their comrades or of being directed or observed by their superior officers.

Now in the difficult days of peace there are isolated posts in all corners of the world being attacked, consolidated, and held by small parties of our old Battalion. The qualities of endurance, loyalty, steadfastness, and self-sacrifice which enabled the Battalion to work together as one man during the fighting in France, will still weld together the detached units and maintain the Battalion as a living force in the world for all time. Should it ever be found necessary to re-form the Battalion as a military unit, the King's Colour will, in the words of Gen. Morgan, previously quoted, "form a rallying-point round which will again be displayed that lofty sense of duty, comradeship, and devotion which has ever distinguished the 8th Battalion The Queen's Own (Royal West Kent Regiment)."

APPENDIX

The following Honours and Awards were gained for service with the Battalion :—

(A) OFFICERS

Victoria Cross

Lieut. D. J. Dean.

Bar to the Distinguished Service Order

Lieut.-Col. H. J. Wenyon, D.S.O.

The Distinguished Service Order

Capt. G. A. McLarty (R.A.M.C.).
Maj. A. H. Pullman.
Col. E. Vansittart.
Capt. H. J. Wenyon.
Maj. N. I. Whitty.

Bar to the Military Cross

Lieut. E. G. Brown, M.C.
2nd Lieut. D. W. F. Tanner, M.C.

The Military Cross

Capt. J. N. Arnaud.
Capt. R. P. Baker.
2nd Lieut. H. W. Beattie.
Lieut. G. S. Bowen.
2nd Lieut. E. G. Brown.
Maj. H. S. Brown.
2nd Lieut. H. Cambrook.
2nd Lieut. F. Carville.
2nd Lieut. J. B. Cryer.
2nd Lieut. C. Trenchard-Davis.
Lieut. E. O. Goulden.
Lieut. W. D. de P. Hall (att. T.M.B.).
2nd Lieut. F. J. Janaway.
Lieut. R. N. Killick, D.C.M.
2nd Lieut. C. P. H. Manley.
Capt. J. C. Orchardson.

2nd Lieut. K. Pfeuffar.
2nd Lieut. D. T. Piggott.
Rev. G. D. Barlow-Poole, C.F.
Capt. A. J. Porter.
2nd Lieut. A. Roscoe.
Capt. J. H. Selfe.
2nd Lieut. B. Stephens.
2nd Lieut. D. W. F. Tanner.
2nd Lieut. P. H. Tyler.
Lieut. W. K. Tillie.
Capt. S. G. Thompson.
2nd Lieut. G. A. E. Wallis.

MENTIONED IN DESPATCHES (3 TIMES)

Lieut.-Col. H. J. Wenyon.

MENTIONED IN DESPATCHES (TWICE)

Lieut.-Col. N. I. Whitty.

MENTIONED IN DESPATCHES

Capt. J. N. Arnaud.
2nd. Lieut. V. G. Don.
Lieut. and Quartermaster H. Evans.
Lieut. W. D. de P. Hall (att. T.M.B.).
Capt. G. A. McLarty.
Lieut. H. G. Michie.
Capt. C. de C. Middleton.
Maj. A. H. Pullman.
Capt. W. K. Tillie.
Col. E. Vansittart.
Lieut. and Quartermaster J. R. Wood.

AWARDED BY THE PRESIDENT OF THE FRENCH REPUBLIC:

CROIX DE GUERRE

2nd Lieut. W. E. W. Thomas.

AWARDED BY THE KING OF THE BELGIANS:

ORDRE DE LA COURONNE AND CROIX DE GUERRE

Lieut.-Col. H. J. Wenyon.

(B) OTHER RANKS

BAR TO THE DISTINGUISHED CONDUCT MEDAL

Cpl. C. Andrews, D.C.M.
Regt.-Sergt.-Maj. R. Rankin, D.C.M.

The Distinguished Conduct Medal

L.-Cpl. C. Andrews.
Sergt. E. H. Cherriman.
L.-Cpl. A. Cook.
Coy.-Sergt.-Maj. W. Dunk.
Cpl. C. Gallagher (att. T.M.B.).
Coy.-Sergt.-Maj. F. Greenaway.
Coy.-Sergt.-Maj. F. Gutteridge.
Coy.-Sergt.-Maj.W.H.Mills.
Coy.-Sergt.-Maj. W. Mitchell.
Pte. G. Neal.
L.-Cpl. G. T. Ogilvie (att. T.M.B.).
Sergt. A. J. Purfield.
Coy.-Sergt.-Maj. R. Rankin.
Sergt. H. T. Skipper.
Pte. A. E. Smith.
L.-Cpl. H. Stoneham.
Sergt. H. W. Terry.
Sergt. E. C. Vanner.
L.-Cpl. A. L. Woolmore.

Bar to the Military Medal

Pte. E. Bowes, M.M.
Cpl. E. H. Cherriman, M.M.
L.-Cpl. F. Laing, M.M.
Pte. A. Yates, M.M.

The Military Medal

Sergt. R. Alderman.
Pte. J. H. Allchin.
Sergt. H. C. Allen.
Pte. H. G. Anderson.
L.-Cpl. F. A. Apps.
Pte. J. Baker.
Cpl. A. J. Bates.
Cpl. J. J. Bellamy.
L.-Cpl. H. Binks.
L.-Cpl. J. C. Bolton.
Pte. E. Bowes.
L.-Cpl. G. S. B. Boylan.
Pte. J. E. Bradley.
L.-Cpl. C. R. Brown.
L.-Cpl. S. Browne.
Sergt. G. Buck.
Pte. W. J. Bush.
Pte. J. W. Calow.
Pte. F. A. Cannon.
Sergt. H. A. Cater.
Pte. E. H. Cherriman.
Pte. D. J. Deacon.
L.-Cpl. A. Dent.
Coy.-Sergt.-Maj. C. R. Dyer.
Pte. A. J. Eagleton.
L.-Cpl. J. East.
Pte. A. J. Everist.
Cpl. J. Eversfield.
Pte. F. Fever.

Cpl. A. French.
Pte. H. A. Frewer.
Sergt. L. Green.
Pte. W. G. Goodwin.
L.-Cpl. F. Holford.
Cpl. W. Jasper.
Pte. E. J. Judd.
Sergt. E. G. Kirby.
L.-Cpl. F. Laing.
Pte. H. H. Lang.
Sergt. G. Laskey.
Pte. E. G. Luckett.
Pte. A. W. Miller.
Sergt. W. Mitchell.
Pte. R. Morris.
Pte. W. W. Paine.
Cpl. T. Pannett (att. T.M.B.).
Pte. F. A. Pierce.
Pte. R. T. Roper.
Pte. H. Rumens.
Sergt. F. Sadler.
Pte. J. D. H. Scripps.
L.-Cpl. T. G. Shewry.
Cpl. G. W. Simmonds.
Pte. R. G. Simmons.
Pte. P. Skerry.
Sergt. C. A. Taylor.
L.-Sergt. R. Taylor.
Pte. H. Tipler.
Sergt. E. Turner.
Pte. A. H. Tyrrell.
Pte. R. Turk (att. T.M.B.).
Cpl. H. G. Warford.
Pte. A. W. Watson.
Pte. S. G. Wells.
Sergt. A. W. Westbury.
Pte. J. H. Wheeler.
Sergt. C. D. Williams.
Pte. J. Winder.
Cpl. C. W. Woodrow.
Cpl. G. Wookey.
Pte. A. Yates.

The Meritorious Service Medal

Sergt. C. Austin.
Sergt. E. Bailey.
Sergt. W. C. Carr.
Regt.-Quartermaster-Sergt. F. A. Ifould.
Sergt. G. G. Shepherd.
Sergt. H. R. Swain.
Cpl. H. R. Verrall.
Sergt. W. Wood.

Mentioned in Despatches

Pte. J. W. Calow.
Pte. A. T. Canning.
Pte. F. Greenhalf.
Coy.-Sergt.-Maj. F. Greenaway.
Coy.-Sergt.-Maj. F. Gutteridge.

MENTIONED IN DESPATCHES—*Continued*

Coy.-Sergt.-Maj. G. D. Jarrett.
Pte. P. C. Jeeps.
Cpl. G. Knell.
Pte. E. Knowles.
Pte. A. Scarratt.
Pte. A. W. Watson.
Pte. J. H. Wilks.

AWARDED BY THE PRESIDENT OF THE FRENCH REPUBLIC:

CROIX DE GUERRE

Pte. F. A. Cannon.

MEDAILLE D'HONNEUR AVEC GLAIVES EN BRONZE

Cpl. H. R. Verrall.

AWARDED BY THE KING OF THE BELGIANS:

CROIX DE GUERRE

L.-Cpl. C. Andrews.

DÉCORATION MILITAIRE

Cpl. A. French.

CAMBRAI TO MAUBEUGE.

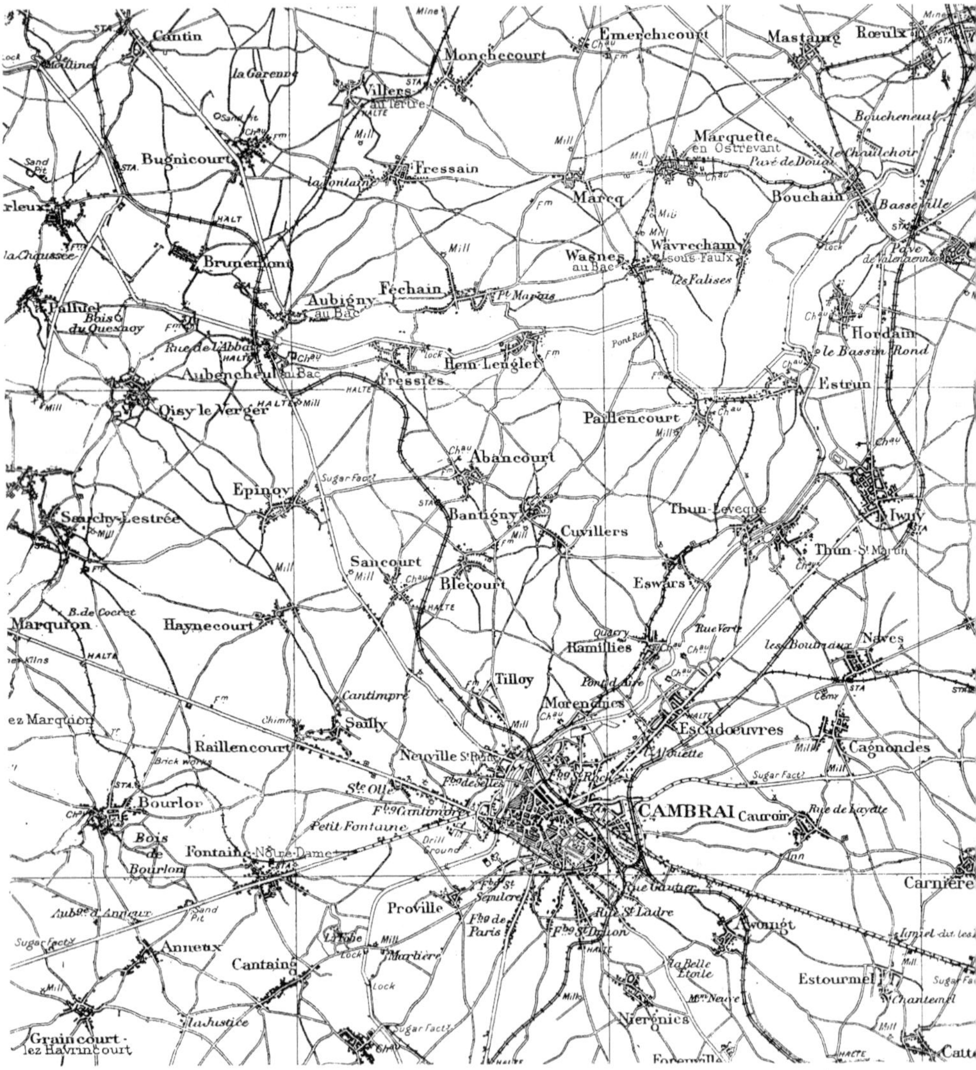

SCALE 1 : 100,000.

0 1 2 3 4 5 6 7 8 9 10 miles.

To illustrate the Final Advance. Phase IX.

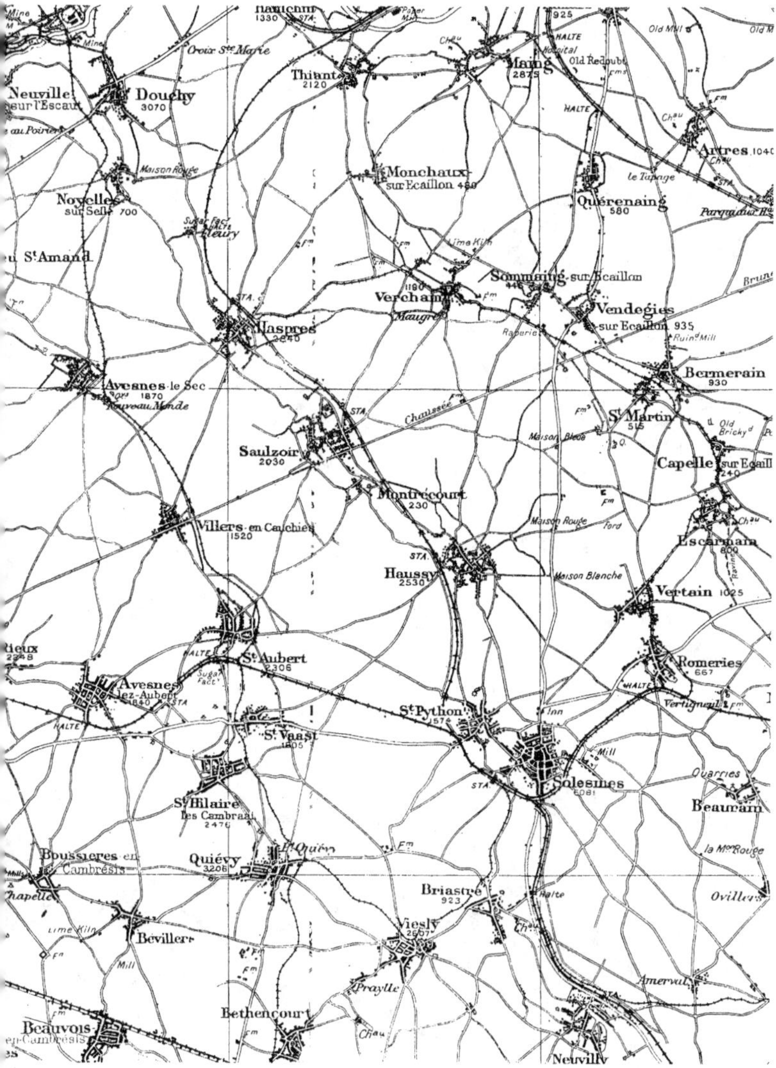
Douchy
3070
Neuville sur l'Escaut
Thiant
2120
Maing
2875
Old Redoubt
Artres
Croix Ste Marie
Maison Rouge
Monchaux-sur Ecaillon
Noyelles sur Selle
700
Quérenaing
580
le Tapage
Lime Kiln
Sommaing sur Ecaillon
Verchain
Maugré
Haspres
2640
Vendegies sur Ecaillon
935
Ruind Mill
Raperie
Bermerain
930
Avesnes le Sec
1870
Nouveau Monde
Chaussée
St Martin
515
Maison Bleue
Saulzoir
2030
Capelle sur Ecaillon
Montrécourt
230
Maison Rouge
Villers en Cauchies
1520
Escarmain
Haussy
2530
Maison Blanche
Vertain
1025
St Aubert
2306
Romeries
667
Avesnes lez-Aubert
1840
Vertigneul
St Python
St Vaast
1805
Solesmes
Quarries
St Hilaire les Cambrai
2476
Beaurain
la Mon Rouge
Boussieres en Cambresis
Quiévy
3206
Briastre
923
Ovillers
Viesly
2607
Bevillers
Lime Kiln
Mill
Prayle
Amerval
Bethencourt
Beauvois en Cambresis
Neuvilly

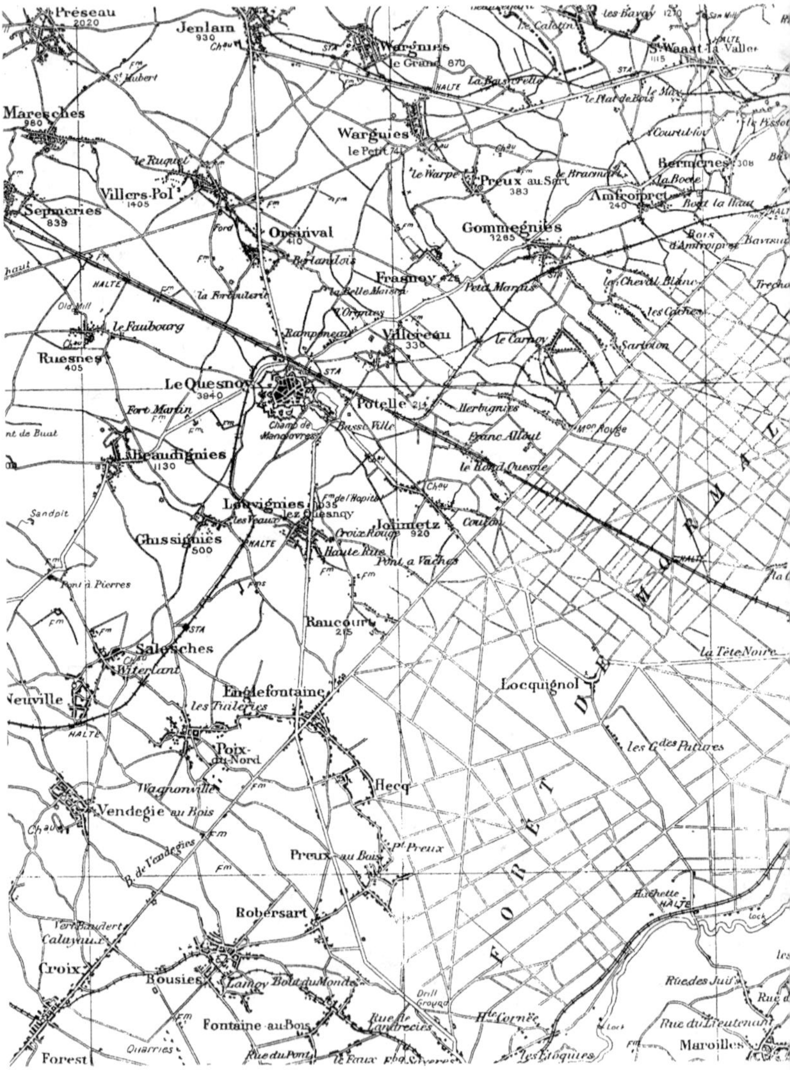
Préseau
Jenlain
Wargnies le Grand
St Waast la Vallée
Mareseches
Wargnies le Petit
Villers-Pol
Preux au Sart
Bermeries
Amfroipret
Sepmeries
Orsinval
Gommegnies
Frasnoy
Villereau
le Faubourg
Ruesnes
Le Quesnoy
Potelle
Fort Martin
Beaudignies
Louvignies lez Quesnoy
Ghissignies
Jolimetz
Croix Rouge
Haute Rue
Pont a Vaches
Raucourt
Salesches
Englefontaine
Locquignol
Neuville
les Tuileries
Poix-du-Nord
Hecq
Vendegie au Bois
Preux au Bois
Robersart
Croix
Bousies
Fontaine au Bois
Forest
Maroilles
Rue des Juifs
FORET DE MORMAL

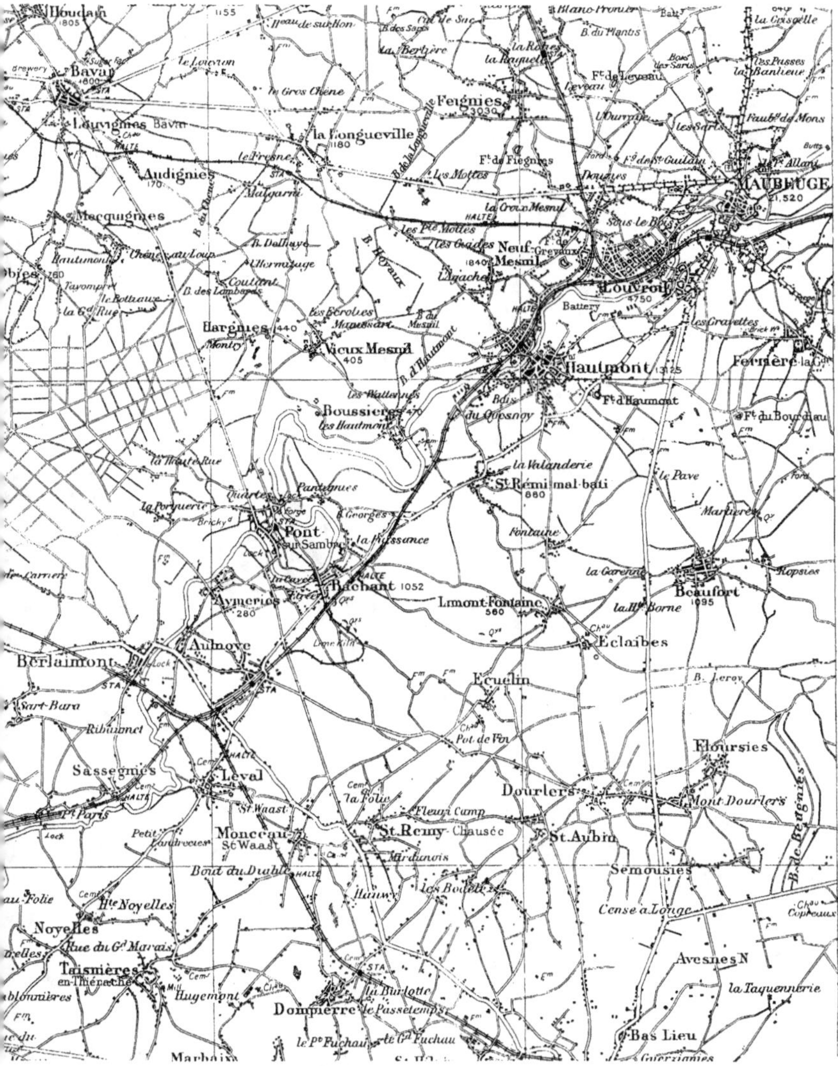

[Reproduced by permission of the Geographical Section, General Staff.

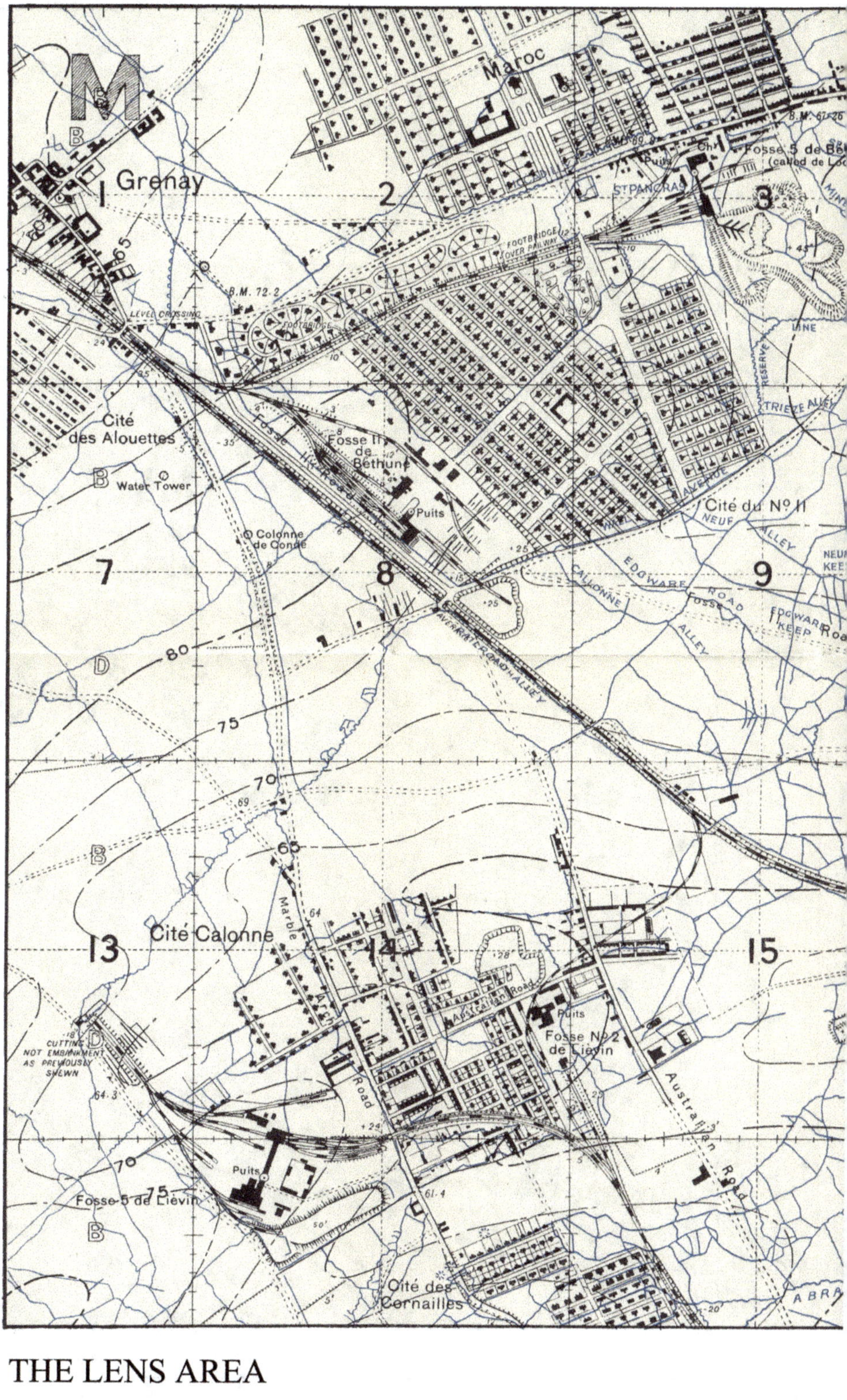

THE LENS AREA

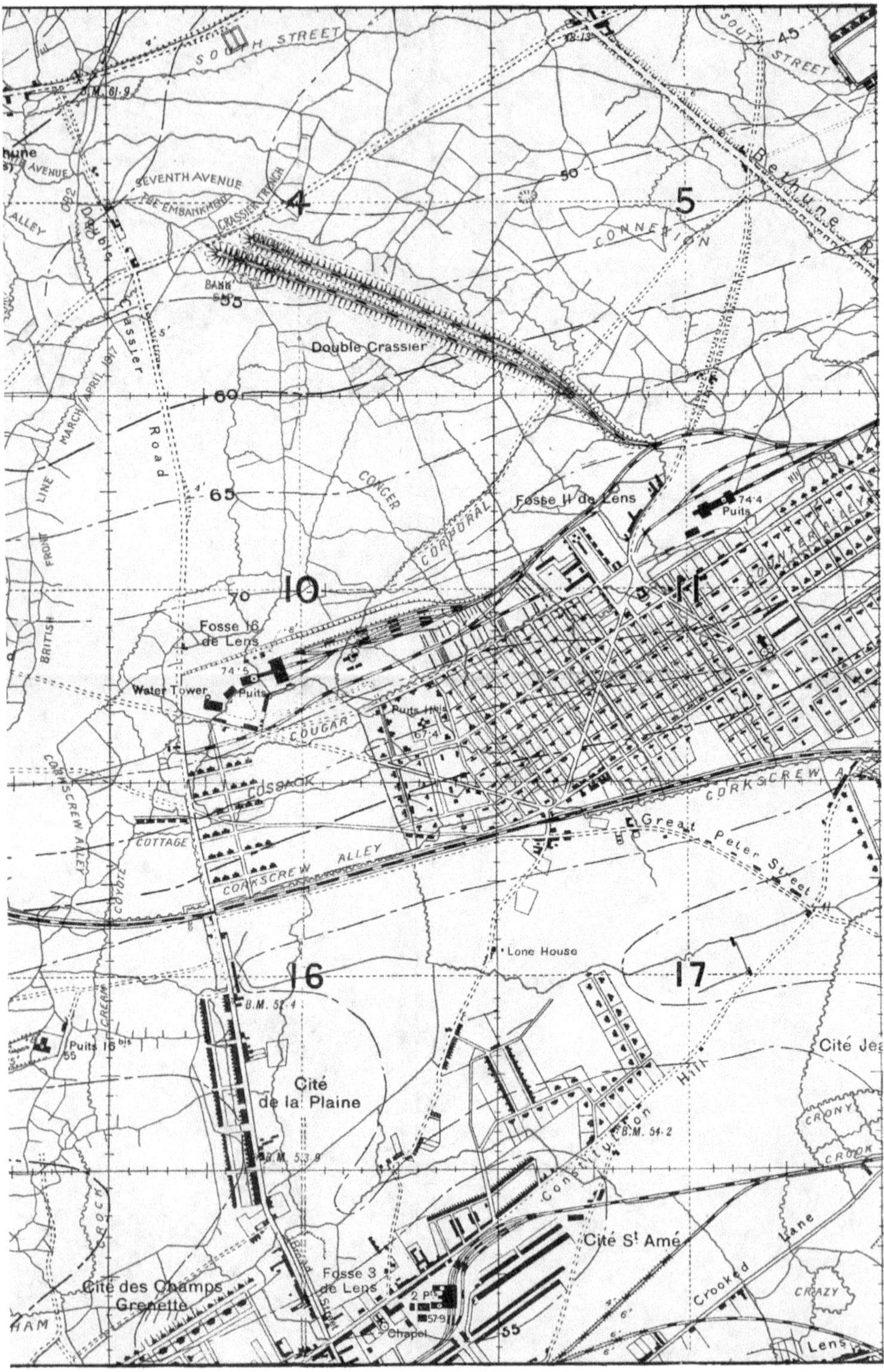

SCALE 1 : 10,000

0 — ¼ — ½ — ¾ — 1 mile.

500 — 1,000 — 1,500 yards.

To illustrate Phases IV and VIII.

THE LENS AREA

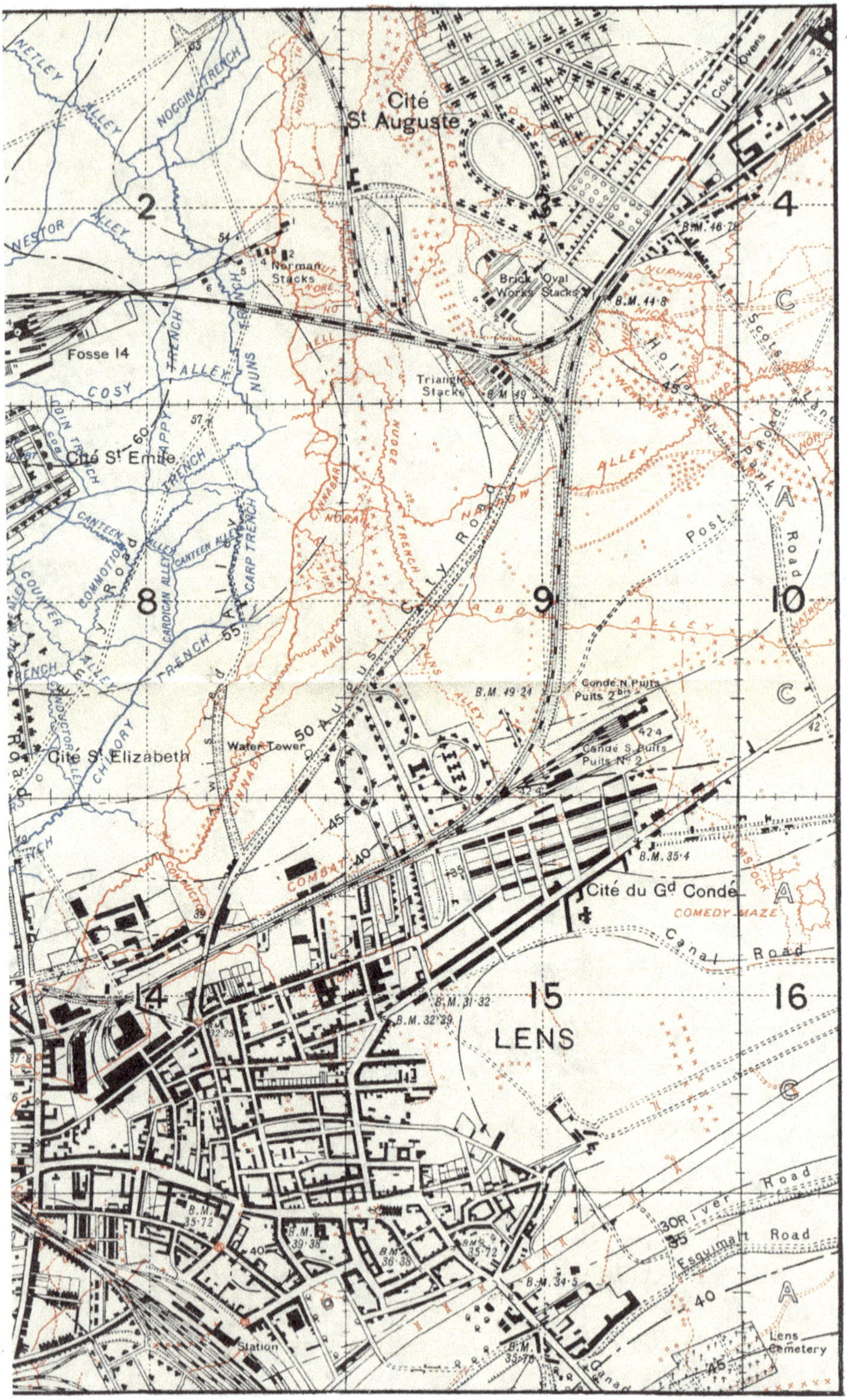

Scale 1:10,000

0 — ¼ — ½ — ¾ — 1 mile

500 — 1,000 — 1,500 yards.

To illustrate Phases IV and VIII.

THE YPRES SALIENT

Scale 1: 10,000

0 — 500 — 1,000 — 1,500 — 1 mile — yards

To illustrate Phases II and V.